The
LAST FISH
TALE

Also by

MARK KURLANSKY

To the people of Gloucester and Newlyn,

to their fishermen and their families,

and to the memory of my friend Harold Bell,

who was a delight to talk to and made

the great city he loved even better

The

LAST FISH
TALE

. . .

The Fate of the Atlantic
and our Disappearing
Fisheries

MARK
KURLANSKY

Jonathan Cape
London

Published by Jonathan Cape 2008

2 4 6 8 10 9 7 5 3 1

First published in Great Britain in 2008 by
Jonathan Cape
Random House, 20 Vauxhall Bridge Road,
London SW1V 2SA

www.rbooks.co.uk

Addresses for companies within The Random House Group Limited can be found at:
www.randomhouse.co.uk/offices.htm

The Random House Group Limited Reg. No. 954009

A CIP catalogue record for this book is available from the British Library

ISBN 9780224082457 (Hardback)
ISBN 9780224085717 (Trade paperback)

Grateful acknowledgment is made to the following for permission
to reprint previously published material:

Harcourt, Inc.: Five-line excerpt from "The Dry Salvages" in *Four Quartets* by
T. S. Eliot, copyright © 1941 by T. S. Eliot and copyright renewed 1969
by Esme Valerie Eliot. Reprinted by permission of Harcourt, Inc.

New Directions Publishing Corporation: Excerpts from "Maximus, to Gloucester,
Sunday, July 19" and from "Moonset, Gloucester, December 1, 1957, 1:58 AM"
from *Selected Writings of Charles Olson*, copyright © 1951, 1966 by Charles Olson.
Reprinted by permission of New Directions Publishing Corporation.

University of California Press: Excerpt from *Selected Letters* by Charles Olson,
edited by George F. Butterick, copyright © 1983 by The Regents of the
University of California; excerpt from *Selected Letters* by Charles Olson,
edited by Ralph Maud, copyright © 2000 by The Regents of the
University of California. Reprinted by permission.

Pen-and-ink drawings by the author.

FRONTISPIECE: *Baiting a trawl on the schooner* Corinthian. *Photo by Edwin H. Cooper,
courtesy of the Cape Ann Historical Association, Gloucester, Massachusetts.*

Book design by Barbara M. Bachman

The Random House Group Limited supports The Forest Stewardship Council (FSC), the
leading international forest certification organisation. All our titles that are printed on
Greenpeace approved FSC certified paper carry the FSC logo. Our paper
procurement policy can be found at www.rbooks.co.uk/environment

Printed and bound in Germany by
GGP Media GmbH, Pößneck

A fisherman is not a successful man

he is not a famous man he is not a man

of power, these are the damned by God.

—CHARLES OLSON,
"MAXIMUS TO GLOUCESTER,"
SUNDAY, JULY 19, 1960

Contents

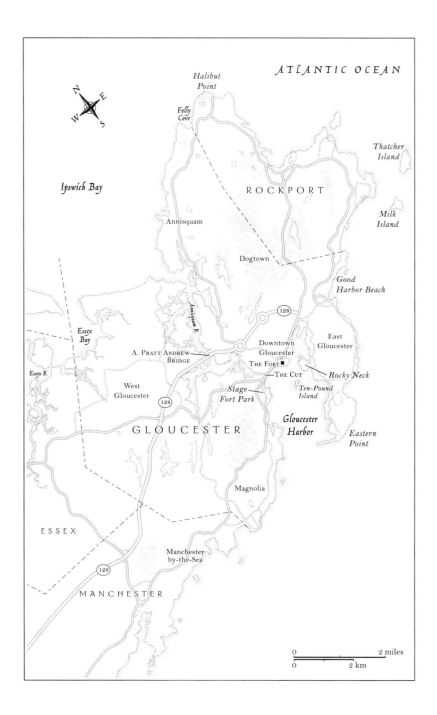

*Halibut
Point*

*Folly
Cove*

ATLANTIC OCEAN

Ipswich Bay

ROCKPORT

*Thatcher
Island*

Annisquam

*Milk
Island*

Dogtown

*Good
Harbor Beach*

128

*Essex
Bay*

Downtown
Gloucester

East
Gloucester

A. PRATT ANDREW
BRIDGE

THE FORT ■

Essex R.

THE CUT

Rocky Neck

West
Gloucester

128

*Stage
Fort Park*

*Ten-Pound
Island*

GLOUCESTER

*Gloucester
Harbor*

*Eastern
Point*

Magnolia

ESSEX

Manchester
by-the-Sea

128

MANCHESTER

0   2 miles

0   2 km

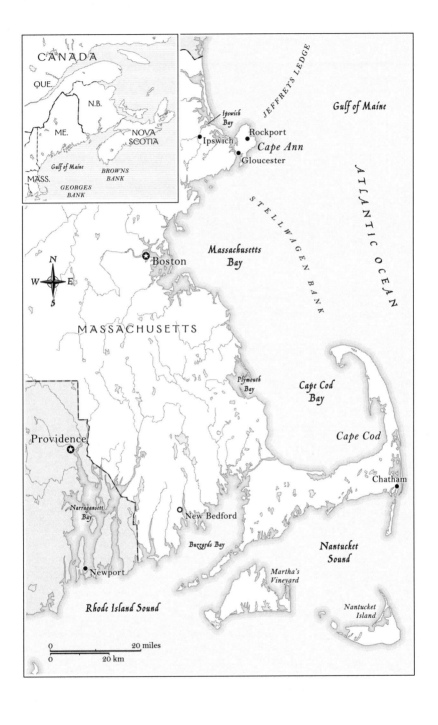

Preface

THE REAL WORLD AND ITS FUTURE

...

Many of us strive to keep our bodies supple, but we do not have much success with our minds. I remember that I only just managed to learn the elements of wireless telegraphy during the last of my teens; but the boys who followed me knew it all. Not many older men will know about inductance; or appreciate the curvature of space; and most of us will close our ears to the cogent argument that only the ideal world is real.

– Michael Graham, *The Fish Gate*, 1943

THOUGH THERE SEEMS TO BE SOME DEBATE ABOUT WHICH IS THE better company, I have long been concerned about both fish and fishermen. This is only one of several reasons why fish farming offers little in the way of a solution to the fishing crisis. The struggle for the survival of fisheries is not only biological but cultural. No country better illustrates this than Britain for if this island nation loses its fisheries, it will have lost its cultural heritage. Without commercial fishing, Britain will no longer be the same place and the British will no longer be the same people.

The Fish Gate, a brilliant study of the crisis in British fisheries, "The first readable book that has ever been written on the subject" according to one reviewer in the early 1940s, by Michael Graham, a scientist who directed the Lowestoft laboratory on the now dying North Sea, is only one piece of evidence for the proposition that the British were in the forefront of studying the dilemma of overfishing, and also in the forefront of overfishing itself. The British prefer not to take the credit for either and, like the Americans and the Canadians, tend to see the Spanish or the Russians or the Japanese as the principal culprits.

Fishermen have a competitive nature. In hard times British fishermen do what they can to defend themselves against the outrages of the Japanese or the Spanish. But, in addition, the English denounce the Scots, the West Country fishermen accuse their North Sea colleagues, and Newlyn fishermen have a history of protesting against the ways of Brixham. And so, not surprisingly, Massachusetts fishermen have never spoken particularly well about their brothers across the sea.

But fishermen have also been among the first to take an ecological view of their crisis – that is, to understand that it is all the same crisis, that the survival of Massachusetts and the survival of Cornwall are intimately connected, that even while scientists point out that most fish stocks are separate, the fate of the Massachusetts Bay is connected to the fate of the Gulf of Maine and that this has everything to do with the future of the North Sea and the Irish Box. If Gloucester, Massachusetts cannot survive as a fishing port, how can Newlyn hope to survive, and then what is the fate of the British?

To the world of Atlantic fishing, Gloucester, Massachusetts is an iconic place. This is why, when Newlyn decided to build a monument to their thousands lost at sea, the first image proposed was the Gloucester monument to its ten thousand gone. For one thing, the two cities have in common the pain of their losses.

Gloucester, were it not for this unity of fishing ports, would be by now completely severed from England. True it was originally settled by English fishing interests in the West Country (though not from Gloucestershire). True too, the place names around Gloucester, such as Penzance and Land's End, tell of the origins of these seventeenth-century settlers. But today's Gloucester has a population that is an amalgam of maritime Atlantic peoples. Far more have roots in Sicily, the Azores, Ireland by way of Newfoundland, and Scandinavia, than the handful of old Gloucester families such as the Tarrs and the Wonsons who can trace their roots to England.

Long after all colonial and commercial ties vanished, Gloucester fishermen followed with interest the activities and prosperity of British fishermen. After all, for much of the twentieth century most of them worked on vessels of British design – stern trawlers. Dragging nets over the stern to fish the bottom of the sea was not a British idea. It is generally believed that the Dutch did it first in the North Sea. But the British developed and perfected stern trawling, known in New England as bottom dragging.

With the development of the stern trawler Britain took a leading role in fishing, in overfishing, and in discussing the problem. Unfortunately little was done to reverse the damage to fish stocks from the new invention. As fish became more scarce, the stern trawlers moved from the southern North Sea to the northern North Sea and ever further afield, looking for ever more fish to feed the tremendous capacity of the modern British trawler. Commissions were formed, studies were made, and nothing was changed.

When the first British-type stern trawlers arrived in Gloucester at the beginning of the twentieth century, many in the local fishing industry protested because they recognised that the British experience would now be their fate too. In 1911 the small local newspaper, the *Gloucester Daily Times*, devoted whole pages to outlining the inability of the British over many years to deal with the problems caused by this new type of fishing gear. The newspaper wrote of the Anglo-French convention of 1839, two years after beam trawling began in Great Britain. Both the French and the British had agreed to restrict the practice, especially in those waters traditionally fished for mid-water fish such as herring, pilchards and mackerel – which is in fact a large part of the British fishing grounds. But the destruction continued and the British government began a regular launching of commissions of inquiry. These were dominated by the biologist Thomas Huxley who seemed to be convinced that it was scientifically impossible for human beings to deplete fish stocks.

But fishermen, constantly observing diminishing fish populations, kept demanding further investigation. In 1883, as the Massachusetts article notes with frustration, Britain was on its third Royal commission devoted to the problem, the so-called Dalhousie commission, after William, Earl of Dalhousie who presided. This commission appointed a Professor MacIntosh of St Andrews University in Scotland to study beam trawlers. After 17 days of hearings, held during a three-month period, MacIntosh reported to parliament:

After carefully considering the whole evidence, upon the question of the decrease of fish, we are of the opinion that as regards territorial waters, the principal fisheries carried on in the inshore waters of the North Sea are the haddock, whiting, flat fish, etc. On many fishing grounds from the Moray Firth to Grimsby there has been a falling off of the takes of flat fish, both as regards quantity and quality. There has also been a decrease in the catch of haddock in certain places, chiefly in bays and estuaries...the fishermen were almost unanimous in stating that the decrease of haddock and flat fish had been contemporaneous with trawling and that it had become more marked since the introduction of beam trawling.

The Dalhousie commission further reported that, due to the overfishing and killing of young fish by trawling, the North Sea had become "exhausted."

And yet, as the *Gloucester Times* pointed out, over in Massachusetts almost thirty years later, these practices were still continuing. They – and even more destructive kinds of trawling – still continue today, though less so in the North Sea since it is nearly dead. In 1883 in Scotland most of the fishermen used hook and line and angrily protested against the new gear, though they were powerless to stop it. In 1900 a select committee appointed by the House of Commons again warned that the North Sea was becoming seriously depleted. By the time of the 1904 select committee, British fishing had moved far beyond the little beam trawlers to far larger beams, facilitated by more powerful engines and otter trawls that dragged a net off the stern. The capacity of British fishing had been increased by many times its size at the time of the 1883 warning.

In 1911 in Gloucester, where most fishermen were still using lines, the newspaper warned that the same thing could happen there. It has.

The *Gloucester Times* stated that:

The failure to secure remedial legislation in Great Britain has been ascribed to the fact that 75 percent of all otter trawlers were owned and operated by British subjects many of whom sell their catch in Germany, France, Spain, Portugal, and other European countries, who would have retaliated had Great Britain closed or restricted her markets to their trawlers. This together with the ever increasing influence of the steam trawling interests probably accounts for the apparent reluctance of Parliament of taking a decided stand on the controversy. And with the ever increasing number of steam trawlers has come a corresponding decrease in catch.

And this explains why such destructive practices are still permitted in both the European Union and the United States, and, for that matter, the rest of the world. It is particularly remarkable in an age in which people will pay more for line caught fish.

And so these remaining fishing places share a common destiny, and just as the people of Massachusetts kept a worried eye on British fisheries, with the English coast of the North Sea dying and the ports of Cornwall and Devon being rapidly overtaken by tourism, the survival of Gloucester, Massachusetts is an urgent British issue.

POLE WALKERS

. . .

Father Sea
Who comes to the skirt
of the City

—CHARLES OLSON, FROM *THE MAXIMUS POEMS*, 1968

BURLY, BARREL-CHESTED FISHERMEN WERE WEARING THEIR favorite dresses, hairy tufts blooming out from the tops of their bodices. There was also an aviator, a brawny nun, a pirate, a gladiator, some overly made-up floozies in need of a better shave, a fedora-sporting gangster, the Jolly Green Giant, and, of course, Dorothy, his white blouse and blue-checked jumper stretched across broad shoulders, braided locks hanging down. The mayor was there, and the lieutenant governor, a carefully groomed, slim, fair-haired woman looking a little frightened. This was, after all, one of the big events of the year and one of the most Gloucester. Gloucester is a town, officially a city, with such a strong sense of itself that the town name frequently is used as an adjective—it's a very Gloucester way of speaking.

The event was the Giambanco sisters' party for the pole walkers on the Sunday morning of the St. Peter's festival. Symbolic acts endure and traditions live on when the metaphor is exactly right. That is the principal explanation for why, on the last weekend of every June, dozens of Gloucester men take a boat out to an offshore platform and walk a forty-foot pole covered with a thick, gloppy cushion of grease, try to grab the flag at the end; and whether they succeed or fail, fall a dangerous two or three stories, depending on the tide, to the frigid June sea below. Because the fall into the sea is inevitable and the chances of injury fair, pole walking is very Gloucester.

The tradition, like many of the pole walkers, has its origin in Sicily. More than half of the Gloucester fishing fleet comes from Sicily or descends from Sicilians. Originally, in contests that trace back to fourteenth-century Sicily, the greased pole was vertical and was intended to be a test of mast-climbing skill.

In Gloucester, the pole-walking event is part of the St. Peter's Fiesta, held on June 29 in Rome since the fourth century, but held in Gloucester on whatever weekend falls closest to the 29th.

Peter, a lake fisherman turned disciple of Christ, is the patron saint of fishermen. He is also the patron saint of shipbuilders and net makers and also of stonemasons, bridge builders, cobblers, and locksmiths—a blue-collar maritime saint, the perfect patron for Gloucester.

The festival came to Gloucester in 1926 when a Sicilian fisherman, Salvatore Favazza, had a statue of St. Peter made in Boston and brought it to Gloucester. Gloucester Sicilians welcomed the statue with a *novena*, a nine-day religious ritual. They mounted the statue on a platform and carried it around the winding streets of the waterfront neighborhood called "the Fort," one of the oldest in Gloucester. An actual fort during the American Revolution and the War of 1812, in the twentieth century this jut of land where fish-processing plants operated became an almost entirely Sicilian neighborhood, with the notable exception of the home of Charles Olson, Gloucester's celebrated poet. During the procession through the Fort, the crowd would shout *"Viva San Pietro!"* and the fishermen would shout back in Sicilian dialect, *"Me chi samiou, duté muté"* Are we mute? Shout it louder. Then the crowd would respond with an even louder, *"Viva San Pietro!"*

Through boom years and crisis, the Sicilian fishermen of Gloucester have always set aside money every year for a weekend to honor the saint of fishermen. The festivities still include nine days of prayer, traditional Sicilian songs, the procession of the statue, the Cardinal's tour of the harbor and his blessings, and a public prayer for the fishermen. But pole walking, carnival rides, rowing contests, fireworks, and other events have been added. Although to the purist the major event is still the solemn procession of the statue and other religious rituals, to be honest, the pole walking has become the big event of the weekend.

IT IS GENERALLY RECOGNIZED that to be a successful pole walker a contestant must be tremendously brave, extremely agile, and extraordinarily drunk.

The pole, the thickness of an old schooner mast or a telephone pole, sticks out horizontally from a platform and is covered about six inches deep with a cushion of white industrial grease. They used to use black grease, which looked more industrial. But the pole walkers come out of the water looking cleaner with white grease. Now they go sliding on what appears to be the largest-ever glob of Crisco. As if this were not slippery enough, not without a sense of humor, banana peels are embedded in the greasy muck. The challenge is to walk out forty feet barefoot and remove a flag that is tacked to a stick at the end of the pole.

Because most pole walkers come from families that work in what is widely recognized in the United States and most other countries as the most dangerous job—commercial fishing— they are undeterred by the numerous ways in which pole walking can cause serious injuries, broken ribs being the most common. If the pole walker feels himself slip and struggles to recover, he may end up falling on the pole, smacking his head or chin on it on his way down, crashing against it chest first, or worse, slamming down with a leg on either side. Surviving that, he will most likely belly flop forty feet into the cold ocean—only the hearty swim in Gloucester in June.

The harbor police circle the platform, ready to fish out the fallen if they are too badly injured to swim the two hundred yards to Pavilion Beach, where the cheering crowd and the bronze statue of a fisherman that is the symbol of Gloucester are watching. Pavilion Beach, once the site of an exclusive hotel by the same name, runs along the Boulevard, Gloucester's attempt at a grand thoroughfare. From Memorial Day to Labor Day, the seaward side of the Boulevard is adorned with 180 American flags. Apparently some Gloucester residents find this a bit excessive or maybe just irresistible; they periodically tear a few down. The previous year at least seventy flags were torn down, shredded, or stolen. This year an angry vigilante group was formed to watch over the flags.

Traditionally, there have been two pole-walking competitions. Anyone can walk on Saturday—or, since 1999, Friday and Saturday—but only the winner is allowed to walk on Sunday, when he competes with all the past winners, the superstars of greasy pole walking. When a champion gets too old or too sensible to walk on Sundays anymore, he can designate someone to walk in his place.

Pole-walking champions differ on technique. Some try to make their way gingerly through the grease. Others take a flying start and hope momentum will get them the forty feet to the end of the pole. It would be reasonable to imagine that people who are good at scampering along a greasy pole would be thin, light-footed men, but these competitors are mostly big, brawny, strapping Sicilians. Anthony Saputo said, "It's a sport. It takes skill, courage, balls, and luck." Saputo is called an "international champion" because he won a pole walk in Terrasini, Sicily, in 1985, and then in Gloucester in 1988. Saputo said, "The secret is to step back to get momentum and then it's all God's will. If you are afraid, you are all done."

Many say the water is so cold you don't feel anything until the next day, but often someone is taken to the hospital with broken ribs or other injuries, usually from hitting the pole but also, when low tide makes the fall longer, from the crash into the sea. One thing the more experienced walkers agree on is that if you feel yourself slip, don't fight it; dive clear of the pole so that you don't slam into it on the way down and you hit the water cleanly.

Anthony Matza Giambanco, the brother of the Giambanco sisters, survived an unsuccessful light heavyweight boxing career with a few injuries that reshaped his nose. But the injury

Phil Parmenteri diving for the flag in the Friday competition in 2001.
(PHOTO BY BART A. PISCITELLO)

he finds most memorable is breaking two ribs winning the Saturday pole walk in 1975, after which he walked the pole on Sunday with his ribs taped with strips cut from nylon bathing suits and won the Sunday competition. He won Sunday titles again in 1977, 1978, and 1980. But this year, 2006, still competing at age fifty-one, he said, "I'm married and have five kids. I don't fight the grease anymore."

Matza is called "The Sheriff." The pole walkers, especially the Sunday champions, like to appear in costumes. For a time, he dressed like a sheriff. "But after a while I had to change up. So I started dressing like a woman," he said, sporting beads that coordinated nicely with his purple-flowered shift. Unlike most of the cross-dressed pole walkers, Matza had stylish accessories supplied by his sister, Rosaria, who has a hair salon.

The title The Sheriff dates back three decades, to the time the Italian pole-walking champion Gaetano Carini came to Gloucester to compete in the St. Peter's Fiesta. Because the competition ends when someone gets the flag, the pole walkers had agreed that the first round would be a courtesy round in which no one took the flag. This ensured that everyone would walk the pole at least once. When Gaetano Carini entered their contest, the Gloucester pole walkers explained this rule to him. First they explained it in Italian and then they explained it in Sicilian. But when it came time for Gaetano Carini's first walk of the afternoon, he walked clear to the end of the pole, grabbed the flag, and dove into the sea. The other pole walkers from up on the platform shouted down at him in Italian to put the flag back. But he swam to shore to greet the cheering crowd watching from Pavilion Beach. According to Giambanco and a few other walkers who recall the incident, Gaetano Carini's mistake was that when they shouted down to him, Carini responded with an obscene gesture made with an upraised digit.

Matza Giambanco, the light heavyweight, jumped in and swam after him. He caught up with him on the beach, and in plain view of the crowd, he pummeled the Italian so badly as to send him to the hospital. The horrified spectators wondered why the new champion they were about to cheer was being beaten. Giambanco explained that he did it because he thought the rules should be enforced. That is how he became The Sheriff. Today, locals often repeat the rule, "You can't take the flag on the first round or you get beaten up."

There are many pole-walking legends, such as the year that champion Phil Curcuru was tied up and held in a basement so that he could not compete.

Though Matza was a four-time champion, most pole walkers agree that the all-time best was Salvy Benson, who managed to seize the flag eleven times between 1968 and 1979. This year Salvy Benson was not competing because of a back problem. Pete Frontiero, the son of the 1958 Saturday and 1963 Sunday champion, won nine times, including seven consecutive Sundays between 1987 and 1993. But the pole walkers rarely accord him the standing of Salvy Benson because he did something not quite legitimate and very unGloucester. He studied and practiced.

"That's why he always won," complained Matza Giambanco, who earns his living as an interior painter. "You'd be working. You'd go down to the Boulevard at lunch and eat a hot dog and see him out there practicing. I had to work." Even less acceptable, Frontiero would ignore all the Sunday drinking and eating and spend the day preparing for his walk. Because of him, the pole walkers banned practicing.

The first time Matza pole-walked, his mother, Rosalia Cilluffo Giambanco, wanted to make sure her son and all the other walkers had a good meal beforehand. She would say, "If something happens you'll at least die on a full stomach."

Rosalia, like most of the Sicilians in Gloucester, came from a fishing village near Palermo. Fishing was the family trade. Rosalia had spent her Gloucester days curing and packing fish for the big Gloucester fish companies, standing on concrete floors eight hours a day with a thirty-minute lunch break. Permission had to be asked to go to the women's room.

Rosalia's passion was cooking. She cooked for anyone who came over to the house. Rosalia was celebrated for her fish soups and linguine and clams. She always said, "If I die I'll die happy if I die full." When she was terminally ill she insisted that the pain she was feeling was from hunger and constantly snacked on olives and anchovies to stave off the pangs. She died, as she had hoped, on a full stomach, just after eating.

ROSALIA'S FISH CHOWDER

as remembered by her daughter Sefatia

2 pounds whitefish fillets

2 tablespoons butter

1 cup chopped onion

½ cup diced celery

3 cups diced raw potatoes

3 cups water

1 teaspoon salt

¼ teaspoon pepper

2 cups milk

Cut fillets into bite-size pieces.

In a large saucepan, melt butter. Sauté the onion and celery in the butter until tender. Add potatoes, water, and salt and pepper. Simmer until vegetables are tender, about 20 minutes.

Add fish and cook at low-to-medium heat for about 10 minutes. Add milk and reheat, but do not boil.

EVERY YEAR ON THE Sunday of the walk, the pole walkers would come over for pasta, seafood dishes, and Italian pastries. Fishermen would donate the fish. After Rosalia died, her daughters carried on the tradition. The crowd is fishermen and fishermen families, and in Gloucester politicians swarm fishermen the way seagulls follow fish. So Bruce Tarr, the young and blustery state senator from one of the oldest Cape Ann families, was there, a rare Massachusetts Republican who nurtures his Gloucester base by supporting fishing issues and attempting to speak very bad Italian. The mayor, John Bell, arrived, also an advocate for fishermen and one of the best-liked people in Gloucester. He had recently won his third term as mayor and one of his opponents, Dave Anderson, had said, "John Bell is such a nice man. I didn't want to have to run against him." Bell carried his extra weight in a cuddly way and had that New England knack for always looking casual, as if he would really rather be out boating, which in his case may have been true. He had been a successful businessman who had interrupted his retirement to try running the city he loved.

The portly mayor looked comfortable at the Sicilian party of cross-dressers and pirates. But the lieutenant governor, Kerry Healey, seemed ill at ease. Born in Omaha, raised in Florida, she was a Republican from Beverly, a town which though geographically close to Gloucester is culturally a long distance away. Gloucester's mayor had irritated his Democratic Party by supporting what later turned out to be her disastrous run for governor, which he did because she, too, had promised to be concerned about the plight of fishermen. He introduced her to people at the party as "someone who really cares about Gloucester's fishing future." She clung to the mayor while the half-naked costumed pole walkers completely ignored her. They knew that they, not some state politician, were the stars of the day.

Out from the house came the ample Sefatia Romeo, one of the Giambanco sisters, with her gruff voice and piercing black eyes. One of the most outspoken members of the city council, she was an intimidating woman, and she marched straight to the mayor and the lieutenant governor and said in a voice so stern it could silence Massachusetts politicians: "No politics today." Sefatia wrested away the blushing lieutenant governor, dragging her by the arm like a school principal with a wayward child, over to the unshaven Dorothy and an Indian, while Kerry Healey faced her fate with the bravest smile she could muster.

Sefatia was a tough talker. Asked about President George W. Bush she once said, "I'd like to slap that smile right off his face," and no one in Gloucester would doubt that, given the opportunity, she might do it. But everyone also knew that Sefatia

Romeo was the softest touch in town. If you were to look in one of Gloucester's all-night supermarkets at about two or three o'clock in the morning, you might find Sefatia. Between her job as a patient's advocate at the local hospital and her city council seat, people came to her with problems about their hospital billing, their insurance, and pleas for City Hall. This is why she went to the supermarket in the middle of the night. "Otherwise," she explained, "it would take four hours to do my shopping."

FINALLY A MAN in a dress shouted, "*Me chi samiou, duté muté!*" And the pole walkers responded with a hoarse, "*Viva San Pietro!*" and started filing out of the house, down the driveway, along an antique and slightly dilapidated street of downtown Gloucester toward the waterfront. One of the women came out of the house and said with great excitement, "We found a pastry left. We have a lot of cannolis and other Italian stuff but this an *American* pastry." There was a special reverence in the tone of the word *American* and the remaining guests gathered around to decide how to divide up the small, quivering, fruit-and-gelatin concoction.

The men marched through downtown Gloucester shouting "*Viva San Pietro!*" The words reverberated off the one- and two-story brick buildings of the waterfront.

"*Me chi samiou, duté muté!*"

"*Viva San Pietro!*" After one round of shouts a champion turned and confessed, "That's all the Italian I know."

They carried money on them, and before they got on the boat that would take them to the platform, they would leave most of what they had at the St. Peter's statue because no one wanted them to pay for anything as they moved from bar to bar. There were a lot of bars in downtown Gloucester and not much time. It was The Sheriff's job to keep them moving. He allotted only fifteen minutes for each bar. About to leave the last bar, someone said to The Sheriff, "Let's have one more beer."

"No," came the voice of experience from the older pole walker. "Let's have a shot. Beer won't keep you warm."

They also stopped at numerous Sicilian homes where they were offered chairs in front of the house. A few sat and were given heaping paper plates of pasta with tomato sauce shining with orange grease. With dazzling speed these were consumed and the pole walkers would get up and make room for other champions, and soon they were all done and off to the next house.

Finally, at about 4:30, now bloated on beer and alcohol and pasta, they waddled onto the dock and down the ramp, which was slippery with a little grease from Saturday. One last stop. They turned, picked up their skirts, and relieved themselves from the dock. Then they climbed into the boat that would take them to the platform.

"*Viva San Pietro.*"

"*Me chi samiou, duté muté!*"

"*Viva San Pietro!*"

"Aw, shit," said one champ with a bare chest and sequined tie. "Look at my toe." It was bleeding slightly. "I jammed it getting onto the boat."

"Does it hurt?"

"Hurt? No, but I've been growing these toenails all year. Now I lost one of my claws."

On the way out, little hand-rolled marijuana cigarettes were passed around and the champions sucked in the smoke, just one more ounce of courage, as they tossed lit firecrackers into the sea while Coast Guard and harbor police vessels looked on. These days fishermen live a regulated life in which a rare moment of lawlessness is savored.

As they approached the slippery ladder up to the greasy platform, someone shouted at the man piloting their boat, "Go a little to the left!"

"Port!" came the angry correction from the inebriated champions.

"Jesus Christ," said one champ dressed as a nun, "Fucking guy comes from Gloucester and he says, 'turn left'—doesn't even know port."

Their moment on the pole had arrived. The experienced leapt clear once they started to slide, the younger ones tried to hang on and smashed themselves into the pole. Some seemed to cartwheel through the air. They all ended up in the ocean.

After the flag was taken, they swam back to Pavilion Beach. One dragged himself onto the beach and, still covered with white grease, stood up. A dark-haired Sicilian woman approached him with an accusatory index finger. "Did you hurt yourself? I was watching and it looked like you hurt yourself."

"No, Mom, I'm not hurt."

"Are you sure? You look hurt," said the angry mother, and she dragged the thickly built one-time champion away.

GLOUCESTER, A CITY BUILT by a sea full of fish, is not like any other. It is an old Puritan New England town, one of the oldest, populated by Irishmen, Scandinavians, Jews, Portuguese, and Sicilians—all the people who came here to work in the fishing industry. It is known for the eccentric inventors who came to test their ideas here, for its great painters, sculptors, and poets, its settlers and its adventurers.

It took four hundred years to build this culture, and it could all be lost in a few decades. Fishing and the culture of fishing, an ancient trade and a way of life that has defined coastal towns throughout history, are vanishing from the Atlantic.

Today in Gloucester an old proverb has a new twist. They now say, "If you give a man a fish, you feed him. If you teach a man to fish, he will starve."

Pole-walking platform from Pavillion Beach

The
LAST FISH
TALE

chapter One

THE
FIRST GLOUCESTER STORY

...

From hence doth stretch into the sea the fair headland
Tragabigzanda fronted with three isles called the
Three Turks' Heads.

—JOHN SMITH,
DESCRIPTION OF NEW ENGLAND, 1616

THERE ARE TWO KINDS OF STORIES TOLD IN GLOUCESTER: fish tales and Gloucester stories. A fish tale exaggerates to make things look bigger. It is triumphal. When in the early seventeenth century George Waymouth reported that the cod caught off New England were five feet long with a three-foot circumference, this may have been a fish tale. We don't know. Surely the Reverend Francis Higginson's reports from Salem in 1630 that lions had been seen running wild in Cape Ann, or that the squirrels could fly from tree to tree, were fish tales.

A Gloucester story is just the opposite. It is a story of miserable irony in which things are shown in their worst light, a story with a sad ending.

Often the history of a place begins with the person who named it. But in the case of Gloucester, the story begins with the men who didn't—the ones who tried to name it and failed. The naming of Gloucester is an entire cycle of Gloucester stories.

MAP OF GLOUCESTER HARBOR, "LE BE
A, Place where their ship was anchored. B, Meadows. C, Little Island. (Ten Pou
Island. (Salt Island.) G, Wigwams of the Savages. H, Little River and meadows. (
plain ground, where there are saffrons, nut-trees and vines. (On Eastern Point.)
River. (Brook near Clay Cove.) O, Little Brook coming from meadows. P, A Bro
At Rocky Neck.) R, Sand Beach. (Niles' Beach.) The sea-coast. T, The Sieur de
plain perceiving the savages. The figures probably denote the depth of water in metres.

Champlain's 1606 map of Gloucester Harbor

RT." Drawn by Champlain in 1606.
nd.) D, Rocky Point. (Eastern Point.) E, Rocky Neck. F, Little Rocky
and marsh at Fresh Water Cove.) I, Brook (at Pavilion Beach.) L, Tongue of
here the Cape of Islands turn. (The creek at Little Good Harbor.) N, Little
t Oakes' Cove, Rocky Neck.) Q, Troop of savages coming to surprise them.
court in ambuscade with seven or eight arquebusiers. V, The Sieur de Cham-

The earliest Europeans to arrive at what is today Cape Ann are thought to have been the Vikings, who, according to the written Icelandic legends known as the Sagas, sailed in 1004 down the North American coast from Labrador to Newfoundland to a place they called Vineland. For a long time it was debated whether to believe this story. But in 1961 the remains of eight Viking turf houses dating to the year 1000 were found in a place in Newfoundland known as L'Anse aux Meadows. Where, then, was Vineland? Today many historians believe that it was the coastline of New England, named after the wild grapes that grew there. According to another story, in 1004, Leif Ericson's brother Thorwald landed on Cape Ann and named it Cape of the Cross. But neither the name nor Thorwald went far. Thorwald died on the expedition and those historians who believe the story at all think that he is buried somewhere on Cape Ann. And that is the first Gloucester story.

In 1606, Samuel de Champlain, a French explorer of the coast of Maine, sailed down to Cape Ann, and seeing three islands off its tip—now called Thachers, Milk, and Straitsmouth Islands—he named the peninsula with the great gray granite boulders marking its headlands, the Cape of Three Islands, which even if said in French, *Cap aux Trois Îles*, is not much of a name. He noted that there were actually two rocky headlands and a passage between them, which he sailed through taking depth soundings as he went, thus charting the course that fishermen home from the sea have been using ever since—from Thachers around East Gloucester to Eastern Point, into the harbor between West and East Gloucester. Champlain thought this was an extraordinary harbor, deep and sheltered, with ample mooring space, and he spent three months charting it. He named it *Le Beau Port*, which was a little more poetic than the Cape of Three Islands, but was destined to be no more durable. Being a skilled seaman, he found anchorage in the safest, most leeward cove in the harbor, but that was not to bear his name either. Instead, the sheltered nook is known today as Smith Cove, named after the young English adventurer, thirty-four-year-old Captain John Smith, who arrived eight years later, in 1614.

This Englishman was very different from Champlain. Though they both were prodigious writers, Champlain's writing revealed little about the man or his life. It is not even certain what year he was born. But Smith's writings are very much about himself, full of praise for his own extraordinary deeds. Historians, distrustful of Smith's braggadocio, tended not to believe what he wrote. Only in recent years has it come to be understood that most of his yarns of daredevil adventures are true.

Smith was a self-made man in more than one sense of that phrase. He not only made his own way in the world—though always assisted by his remarkable ability to attract the patronage of the wealthy—but he also used his writings to establish a colorful persona for himself.

When he was still a teenager he went off to the Lowlands, as did many adventurous young Englishmen, to help the Protestant Dutch fight for their freedom from the despotic Roman Catholics of Spain. The savage combat of that war produced many disillusioned young veterans for the new colonies.

But Smith, after three years of fighting the Catholics, did not cross the Atlantic and instead joined the Austrian army to fight the Turks, the true infidels, in Hungary.

According to his immodest but generally accurate journals, Smith showed great cunning and courage and was rapidly promoted to the rank of captain, but then was captured by the Turks, who sent him into slavery in Turkey. There he found himself owned by an aristocratic woman whom he called "the young Charatza Tragabigzanda." According to Smith, her name meant "girl from Trebizond." Smith and his "fair mistress," as he put it, developed some kind of friendship and it seemed a happy time until she gave him to her brother. The exact nature of Smith's relationship with his fair mistress is unclear, but throughout his life he would be rescued and befriended by young women of social standing. Many historians believe that Tragabigzanda had sent him to her brother to learn their language and customs—she had communicated with Smith in Italian—and that her plan was then to marry him. But Smith either did not understand or did not think much of this plan. He murdered her brother and escaped, traveling by horse to the Ukraine, then Poland, and on to Western Europe.

COLONIALISM WAS THE GREAT opportunity for young English adventurers of the early seventeenth century. There were no traditions and few rules, and a resourceful young man could invent as he went along. Smith was personally involved in the two most important British colonies in North America, Virginia and Massachusetts. At the time, Virginia was seen as the more promising and it was the one that attracted sponsors and investors. But Smith believed that New England had better prospects for the future. The reason for this, he argued, was its wealth of fish. Smith maintained that New England fish were a natural resource worth more than gold. "The sea is better than the richest mine known," he wrote.

This stance was surprising in an age of exploration dominated by highborn men who considered themselves above such activities as fishing and bringing fish to market. The traditional source of wealth for the great men of the Age of Exploration was gold.

It was his 1614 voyage that had convinced Smith of the importance of fishing. Having never been a man of affluence, one of the goals of his voyage was to somehow get rich. He had hoped to find gold but could find none. He then tried whaling, but did not encounter a suitable species. The only thing left was fishing, an activity he despised. So he ordered his men to go fishing, while he had himself and a crew lowered in a small boat to begin charting the coast, an activity he had developed a great fondness for in the Chesapeake. The little boat would move into every inlet and measure its nooks and turns, sounding depths so that their value for future ships would be documented.

Smith charted the coastline with great accuracy from the mouth of the Kennebec in Maine to Cape Cod and tried to name everything as he went. The British had simply referred to the area as part of Virginia, and it was Smith who gave it the name

New England. But less successful were some of his other names. The island off Portsmouth, New Hampshire, that he dubbed Smyth's Iles—his creative spelling, farfetched even for its day, would become infamous among latter historians—is today the Isles of Shoals. His name for the granite-studded peninsula, today Cape Ann, was Tragabigzanda, and he called the three is-

Gloucester Harbor, circa 1890
(COURTESY OF BODIN HISTORIC PHOTO)

lands the Three Turks' Heads in remembrance of three Turks whom he beheaded in duels six months before his capture.

Along the way, always looking for a little profit, he picked up fur pelts. He identified and sounded twenty-five "excellent good harbors," but for reasons that are not clear he neglected to sound the harbor on the tip of Tragabigzanda, perhaps be-

cause Champlain had already done so. He did name the points along the tip of the coast including Halibut Point, which still bears the name, though while it is known as an excellent spot for catching striped bass and sometimes bluefish, there is no record of its ever being a place to land halibut.

Smith returned to Europe not only with his charts but also with seven thousand "green cod," or salted fish, and forty thousand stockfish, or dried cod, that his men had landed off of Monhegan Island in southern Maine. England had recently opened trade with Europe and Smith was able to sell the fish in Málaga, where the price was high. Málaga also had a slave market, where he was able to sell twenty-seven locals whom he had lured onto his ship in New England.

The slave sale did not receive wide attention, but the reputed fortune he made on the fish became legendary. Some years later Massachusetts Governor William Bradford heard that it had been 60,000 fish, a fish tale. The fish, the map, the name New England—all made a strong impression back in England. In fact, it was this story combined with Smith's map that convinced the exiled Pilgrims in Holland to start their colony in Smith's New England.

It was not a moment too soon, according to Smith. In the sixteenth and seventeenth centuries, while explorers were searching for gold and spice routes, fortunes were being made on fishing, especially fishing cod. France and England were to fight repeated wars to secure control of North America's fish.

In an age without refrigeration, the fish that was eaten by most people was salt cured. Cod, a large and plentiful fish with white, flaky flesh, a high degree of protein, and almost no fat, was considered the best fish to salt cure. Sixty percent of the fish eaten in Europe was salt-cured cod.

Since mariners did not know how to measure longitude, it was difficult to measure east-to-west progress on Atlantic crossings. Latitude, on the other hand, was easily fixed by the stars and so sailors knew where they were from north to south. So the easiest way to cross the Atlantic was to pick up a wind, fix the latitude, and stay on it, which was called "westing" and "easting." The French wested to Nova Scotia and the English to Newfoundland and Labrador. Only the Basques and the Portuguese could have wested to New England. But they didn't, perhaps because the Basques while pursuing whales had by chance discovered the more northerly cod grounds years earlier.

Europeans had been fixed on fishing in northern latitudes, but by the second decade of the seventeenth century, this was changing. As the fishing moved south, Europeans discovered a basic truth of marine life. While cold waters are richer in nutrients, and therefore have more plentiful and more desirable fish species, these species grow faster and larger in the southern part of their range. The southern part of the range of the Atlantic cod is New England.

In 1602, Bartholomew Gosnold, an Englishman who worked with Smith a few years later in establishing a settlement in Jamestown, Virginia, crossed the Atlantic in search of the by-then old and often disproved theory that there was a westward passage to China. It is extraordinary how much evidence it took to convince Europeans that North America was very large and in the way. Realizing that there was no route to China around Nova Scotia, Gosnold followed the coastline south as far as Narragansett Bay. Along the way he noted not only the plentitude of cod—he said they constantly "pestered" his vessel, which may be another fish tale—but also that the schools swam in more shallow water, closer to shore, and that the fish were far bigger than those to the north. He renamed the peninsula that the Italian, Giovanni da Verazzano, had named Pallavisino after an Italian general. Gosnold called it Cape Cod, and that was one name Smith did not try to change.

The following year fish merchants in Bristol sent out two vessels to confirm Gosnold's findings and their commander, Martin Pring, concurred that the fishing in New England was better than in Newfoundland and that there was an abundance of rocky coastline, ideally suited for laying out salted fish to dry in the sun.

But meanwhile the French in Nova Scotia were also moving down the coast, enticed by impressive catches landed by Basques from St. Jean-de-Luz. Smith wanted to make more voyages to New England. On the second voyage he planned to establish a settlement but failed to reach New England; in his third attempt in 1615 he was captured by French pirates, escaped, returned to England where he settled down and wrote books, and never crossed the Atlantic again.

And so it was an important step for the English when, in 1620, a group of religious fanatics—Smith's map in hand—set sail, not for Virginia as is often said but for "Northern Virginia," for John Smith's New England, and specifically for a place called Cape Cod, where they would build their religious colony and support it with fishing. Because Smith had overlooked Gloucester Harbor, or maybe just because one peninsula was called Cape Cod and the other Tragabigzanda, settlement began in Plymouth, on Cape Cod and not in Gloucester. It was becoming clear that to promote the new lands, having the right name was critical.

As they crossed the Atlantic, there were several things the Pilgrims did not know. One was how to fish. Another was that a monopoly on New England fishing had been decreed that, if enforced, would stop them from fishing. Had the long-argued measure been passed a few months earlier, the Pilgrims might have gone somewhere else, denying the English their first real foothold in New England.

Earlier attempts by the English to settle New England had failed. In 1607, English settlers in Maine had built New England's first transatlantic vessel, which they used to flee their settlement for England, declaring New England uninhabitable. In the first decade there was a great deal of discussion about whether "over-cold" New England was habitable at all.

Curiously, the fact that native peoples were already living in New England, surviving quite well, did not help to resolve the argument. The issue was whether this place was suitable for an Englishman. The English who went to Cape Ann usually wrote admiringly about the physique of the indigenous people. On Cape Ann, the earliest evidence of human settlement dates back 11,000 years to near Ipswich. This ancient people left behind fishing tackle: stone sinkers and weights for nets and bone hooks. They were still there when Champlain arrived, though they had never attained the sophistication of tribes living farther to the south. Champlain described how they burned out logs to make canoes, how they caught cod, how the women dove for lobsters and could catch a hundred pounds in a day's work, and how the women dried lobster and fish to preserve them for winter food without the use of salt. When Champlain first saw Gloucester, it was spotted with summer gardens, as it is today. But the genocide that took centuries in some parts of America was accomplished in coastal Massachusetts within three years.

When Champlain came to the Cape there may have been as many as 3,000 Indians living on the peninsula that they called Annisquam. Neither Champlain nor Smith paid much attention to this name or these people. But in 1616, once the Indians were exposed to Europeans, before there were even European settlements, they were ravaged by an epidemic of what was probably smallpox along the coastline from Maine to Cape Cod. Indians, according to various theories—because they had lived primarily in the subarctic or because they had few domestic animals with whom to share disease—had experienced no exposure to many of the illnesses from which the rest of the world suffered.

By 1619 there were few Indians left in coastal Massachusetts, leaving European settlers the impression that they were moving to a largely empty, perhaps uninhabitable continent. So in 1630, when the Reverend Higginson observed that "the Indians are not able to make use of the one-fourth part of the Land"—which was the standard European explanation during three centuries of why Europeans should take over Indian land—it appeared to be true.

THAT THE FIRST New England settlement in Plymouth nearly starved is not surprising. It is more surprising that the settlement managed to survive. Activities such as fishing and farming require skills that these "brethren" did not possess. They had to send back to England for experts—"merchant adventurers"—to show them how to fish and how to make salt.

But Cape Ann from the beginning had fishermen and fish. It also had chaos. In 1623, while the Plymouth settlers were still struggling to master fishing and assure a food supply, an English fish-trading company, the Dorchester Company, decided to establish a permanent fishing station in New England and by a series of chance navigational decisions ended up settling the far tip of Tragabigzanda. They landed on the west side of the harbor and built a "stage" for curing fish, and they constructed huts nearby in a place called Fishermen's Field, where a much-enjoyed hilly public park by the sea, Stage Fort Park, is now located in Gloucester. This was the first European settlement on Tragabigzanda, from which the settlers filled their ship with cured cod and sold it in Bilbao, taking twenty-four crew and fishermen with them and leaving behind fourteen farmers. This marked the beginning of a long relationship for Gloucester, and later Boston, with the Basque port of Bilbao, which, being close to Gloucester's latitude, was one of the more conveniently located European ports and was the port of an iron-producing area known for chains, anchors, and other iron fittings.

The fourteen farmers who had been left behind spent much of the next months alone in North America, staring anxiously at the horizon until their ship returned with more men. The permanent population rose to thirty-two, including a man in charge of fishing and another in charge of farming.

The hungry brethren of Plymouth were increasingly interested in the fishing activities of Tragabigzanda. Back in England, the Plymouth Company was given a grant to settle in
Cape Ann where they were free to "ffish, fowle, hawke, and
hunt, truck and trade." Anyone who managed to stay for seven
years was given thirty acres of land. There, on the tip of
Tragabigzanda by the Beau Port that Champlain loved, settlers
found a viable location for a fishing community, a considerably
better spot than Plymouth or the swamps of Cape Cod.

The following year, 1624, Smith conferred with Prince
Charles of England, the future Charles I, on his map of New
England. Charles noticed a peninsula in Massachusetts with a
new settlement, a place called Tragabigzanda. What kind of
name was that? The thirty-two fishermen now settled there
had never thought Smith's tales of the young pagan woman
were an appropriate basis for the name of their community.
Charles thought it would be more suitable to name the peninsula after his mother, Queen Ann, and so Cape Ann became the
fifth recorded name for the peninsula.

The residents of Annisquam died off as did the invaders of
Cape of the Cross. The mariners of *Cap aux Trois Îles* didn't
stay, and John Smith's Tragabigzanda was simply inappropri-
ate. There are few stories from the age of exploration of a place
having this much trouble being named. It was a Gloucester
story, and yet Gloucester itself still didn't have its name.

chapter Two

A TALE OF WOE

...

The ship which brought them
we don't even know
Its name, or Master

as they called a Capt
then, except that that
first season, 1623

the fishing, Gloucester
was good: the small ship (the Fellowship*?) sailed*
for Bilbao full

—CHARLES OLSON,
THE MAXIMUS POEMS, 1960

IN MOST OF NORTH AMERICA, THE FIRST EUROPEANS LEFT poetic, hyperbolic descriptions of their new land using words such as Eden and paradise. The settlers of Tragabigzanda left no such writings. They might have noted sweeping beaches and wild, massive rocks, dramatic natural fortresses and dark, thick woodlands, and hills offering vistas far out into a sea that sparkled in the sun and turned black and dangerous without warning. They might have noted that everywhere at unexpected corners, gigantic boulders from a bygone geological era appeared like lumbering monsters obstructing geometric divisions of the land. Or that on wild-growing bushes in the

Sunset from Niles Beach

swampy lands west of the harbor bloomed fragrant white flowers, *Magnolia glauca*, a rare northern magnolia whose only native habitat in Massachusetts is this spot. It stubbornly refuses
to thrive when transplanted.

Cape Ann was formed in a great and lovely mishap. The
black-speckled white rock that covers, shapes, and defines the
cape, that even today stands immovably in backyards, breaking
up sidewalks, determining the path of streets, is an unusually
dense and hard variety of granite. The density is characteristic
of granite normally found deep in the earth. When glaciers
started moving south they carried with them rock churned up

from below the surface. Off the coast of Massachusetts the ice hit a crevice caused by a fault line and hung up there, started to melt, and deposited large amounts of rock. Cape Ann was formed. The fact that it is on a fault line is little remembered today because there has not been an earthquake there since the eighteenth century, when there was not yet a Richter scale. But had there been one, geologists think it might have registered as high as an 8, which is a very dangerous earthquake.

But there was no commentary on any of this. The principal characteristic the first settlers noted about their new home was that it was a great place for fishing. By the time Tragabigzanda was settled, before the Pilgrims learned how to catch and salt fish, Europeans had discovered their waters and barks from French and British ports were hauling large cod from the sea in record catches, sometimes within sight of land.

The Europeans sent over large barks and lowered small two-man rowboats called dories to fish, then gathered the cod on the bark, salted it, and sometimes took it to the shore to be dried on the rocks. Dory fishing is a method rooted in the medieval Basque whale hunt. The Basques had built stone towers along their own coast from which to signal when a whale was seen breaking the surface of the water. Then the whalers would quickly launch long rowing boats and rush out to harpoon the whale. But the Basque whalers, being skilled but unwise, killed off their whaling grounds and then had to load their rowboats in seagoing barks that would take them north to surviving whale grounds, which also turned out to be cod grounds. And so began the dory fishery.

When the bark was loaded with cured fish, it sailed back to

Europe. Once Europeans started fishing off of Cape Ann they had a shorter return voyage than from most of North America. The voyage from Gloucester to Portugal is one of the shortest transatlantic voyages, the only shorter ones being those to the far north, such as the route from Newfoundland to the Bristol channel.

But Cape Ann had an advantage over Newfoundland. Not only were New England cod bigger and easier to catch than the more northern stocks, but the climate was considerably milder in New England, allowing for year-round fishing, whereas in the north there was only a summer fishery. This made an enormous difference. It meant that Cape Ann and almost all of coastal New England were suitable for permanent settlement, whereas the north was suitable only for summer fishing stations, where just a few remained through the long winter. The fact that people were permanently living on the Massachusetts coastline offered the possibility of a broader, more complex economy, one based on fishing but also involving agriculture and manufacturing because the settlers needed building materials, food, clothing, dry goods, and also services and merchants. New England winter fishing established a difference that 150 years later would separate the colonies dependent on England from the colonies with independent economies that wanted to be free of a mother country.

But there was also something unique about the location of Gloucester, of a sheltered harbor for a fishing fleet on the end of Cape Ann. Most fish, and in particular "ground fish"—the New England term for the white-fleshed fish that feed off the bottom of the ocean and comprise most of the value of New

England fishing—do not live in the dark recesses of the ocean's depths. They live in relatively shallow water that sunlight can penetrate. This light plays a role in the synthesis of nutrients, just as sunlight does on dry land. In most of the world, the majority of fish live on the continental shelves. The continental shelf of North America was the rich fishing ground that Europeans initially discovered. But soon they discovered another. Farther out at sea were a series of wide shoals—shallow areas called banks—teeming with fish. This was not a new concept to European fishermen. Cod had been found to be most plentiful in the North Sea on such banks.

But these banks were larger. From Newfoundland to Massachusetts a series of these banks runs offshore one hundred or more miles out at sea. The farthest out, the Flemish Cap, is more than two hundred miles from the nearest ports. The largest of the Canadian banks, Grand Bank, is larger than Newfoundland. The southernmost bank, off of New England, Georges Bank, also partly claimed by Canada, is larger than Massachusetts. The closest port to the two southernmost banks, Browns and Georges Bank, is Gloucester, and that has always ensured its importance as a fishing port. But it is also close to the rich continental shelf of the Gulf of Maine and the bank that lies between Cape Ann and Cape Cod, Stellwagen Bank. Gloucester had easy access to three of the richest fishing grounds in the world.

ALTHOUGH THE DORCHESTER COMPANY was formed in England in 1622 in order to harvest these teeming American seas, it was not until their third year, 1625, that they considered their Cape Ann fishing venture a success. Not long after, they reversed their assessment, giving up on the entire Cape Ann project as a failure. How could a fishing company have failed in a spot proven by history to have some of the best fishing grounds in the world? Fishing has always been a precarious business. One year appears a grand success, the next is a complete failure, and Gloucester's history has been one of periodic crisis.

The competition between the Plymouth Company and the Dorchester Company was ferocious. The Dorchester men seized the Stage from the Plymouth Company and a tiny civil war might have broken out had not the Plymouth leader, Miles Standish, preferred to withdraw and avert an armed conflict. By 1627 the Plymouth Company dissolved. Just when the fishing venture seemed to be paying off for the Dorchester Company, they became overly ambitious and expanded into farming Fisherman's Field. But it was difficult for fishermen to cultivate a field when the fishing and farming seasons were simultaneous; and in the spring, the settlers had to decide whether to plant the field, thereby missing out on some of the best fishing, or let the field go fallow. The little settlement was in a crisis. By 1628 more ships and more settlers arrived and the crisis was over.

But the first Gloucester story of enduring fame did not come until 1635. Until this story circulated, the small fishing settlement at the tip of Cape Ann was little known even by other New England settlers. Then, suddenly, it became a famously sad place. On August 12, 1635, a small boat named the

Watch and Wait—the best Gloucester stories always blend tragedy with absurdity—sailed from Ipswich around the tip of Cape Ann to Marblehead, on the other side. Today, by land, it is a half-hour journey. The vessel carried twenty-three people, including four seamen and two families with a total of ten children.

As the small ship rounded the tip of the peninsula, gale winds crept up. At about ten at night the growing winds split their sails and they anchored for the night. In the early morning hours the wind became so strong that it began tossing the ship and dragging the heavy iron anchor across the rocky bottom. The *Watch and Wait* helplessly skidded over the choppy seas until it smashed into rocks. The two fathers, the Reverend John Avery and Anthony Thacher, who had recently arrived from London, and an Avery son and a Thacher daughter were swept overboard by a wave and landed on a rock. From there they shouted against the thundering winds for the rest of the passengers to join them on the rock. Thacher later described it as a flat rock wedge between two larger rocks. There was space for everyone. Then came another huge wave and this one, rather than washing passengers onto the rock, hurled the entire vessel onto it, splintering it into bits of broken lumber while at the same time sweeping the survivors off of the rocks and into the swirling, white-foamed sea. For about fifteen minutes Thacher struggled in a swirling maelstrom and several times avoided being mangled on the rocks. Then his feet felt the bottom. The water was only up to his chest, the shore a short walk away.

Upon the shore he stopped for a minute to give thanks to

God for being spared. Then he turned back toward the sea to search for the others. But he could find no one. And then climbing from a heap of broken timber came his wife. But the other twenty-one, including their four children, were never found again.

The pain of the Thachers' loss became legend in New England. So great was the sympathy for the Thachers, who had been unknown because they had been in New England for only a few weeks when the tragedy occurred, that the Massachusetts General Court gave Thacher the elongated rock-crusted little island on which their ship had been wrecked. It was named Thacher's Woe and later Thacher Island.

For surviving family and friends there are few tragedies

Wreck of the fishing schooner Ralph Brown, *circa 1926*
(PHOTO BY GARDNER LAMSON, COURTESY OF THE CAPE ANN
HISTORICAL ASSOCIATION, GLOUCESTER, MASSACHUSETTS)

worse than losing a loved one at sea, having him swallowed up by the heaving ocean. Psychiatrists have written many volumes about "the need for closure," the pain of losing someone without a body to bury. Most people who have lost someone they loved at sea never completely recover. Thacher never did. He did not settle at the tragic site that he now owned, but he did remain on the Massachusetts coast, living for a while with his wife in Marblehead, where his rescuers had taken them. There they had another son. The ornate cradle Thacher painstakingly carved for his new child still survives. Later they moved to Yarmouth, a new Cape Cod settlement, where he took on a long list of civic duties, including, poignantly, anchor inspector. He lived to seventy-eight, but neither the Thachers nor many other New Englanders forgot their tragedy. In the years to come there would be many other such Gloucester stories. Poets, from Henry Wadsworth Longfellow to T. S. Eliot, wrote about the tragedy of ships going down off of Gloucester. The sea would take thousands upon thousands of Gloucestermen and the resulting pain shaped the town of Gloucester as much as did the dense, gray-spotted granite or the fine, white-flaked cod.

Eighty years after the tragedy, a Thacher heir sold the unused island to a Gloucester resident to graze his oxen. In 1771, with only ten working lighthouses in the thirteen colonies, Massachusetts Bay Colony proposed the eleventh on Thacher Island. Recalling the still-remembered Thacher tragedy and numerous subsequent shipwrecks, Marblehead shipping merchants, led by John Hancock, who had important investments in shipping, built the lighthouse, financing it with taxes on

Boston shipowners. The 1770s were not a popular time for
taxes in Boston and many refused to pay, but the lighthouse—
in fact, twin stone towers—was built and still today clearly
marks the rocks off the tip of the peninsula. For fishermen returning from sea it is often the first landmark sighted, pointing
the way southwest to safety beyond Eastern Point at the opening of Gloucester Harbor.

Thacher Island from Bass Rocks

chapter Three

THE ISLAND NAMED GLOUCESTER

...

Goodbye red moon
in that color you set
west of the Cut I should imagine
forever Mother

—CHARLES OLSON,
"MOONSET, GLOUCESTER, DECEMBER 1, 1957, 1:58 AM"

MOST OF THE EARLY SETTLERS OF CAPE ANN CAME FROM Wales and the West Country, the part of Britain closest to North America. English merchants in the West Country saw New England as the far western outpost of the lucrative European fish trade. In the first decade, the 1620s, Boston Harbor was seldom visited, while ships regularly landed settlers and supplies on Cape Ann and carried cured cod back to Europe. The first record of a woman's moving there was in 1627.

By the 1630s, West Country merchants were building the prerequisites of a permanent fishery on Cape Ann—stages, boatyards, and taverns. In 1634 the last vestige of explorer romance was removed by the Reverend Richard Blynman, who came from Plymouth with a group of settlers and went to court to establish the tip of Cape Ann, including Champlain's Beau Port, as a "plantation," an officially licensed settlement. They gave it the name Gloucester because a number of the new group were from Gloucester, England, at the mouth of the Severn, which opens to the sea at the Bristol Channel.

It is odd that Gloucester would end up with this name that had so little to do with its history. These Gloucester people, whose choice of name has endured, were not among the original settlers and little is known about them other than that they were the ones who petitioned the court to establish the plantation. But little is known about the original settlers either, or about the first fourteen worried farmers. Settlers left Cape Ann as quickly as they came, mainly because of disputes like the one between the Dorchester and Plymouth Companies. For decades the town had a transient population. In 1651 an entire group, embroiled in a dispute, left Cape Ann and settled in New London, Connecticut. But at the same time, new settlers kept coming. By 1660 half of the seventy-two households that were living in Gloucester in 1650 had left. The constant losing and gaining of population was something else that would become part of the character of Gloucester.

In 1630 some settlers began building homes away from Fishermen's Field in a place that kept the Indian name, Annisquam. This village, still part of Gloucester today, is on Ipswich Bay, where the short but wide Annisquam River opens onto the Atlantic on the opposite, northern side of the peninsula from Gloucester Harbor. The Annisquam was actually a tidal creek and not a river, for on its southern end, where the little streams of the headwaters would be if it were a river, there were only a few yards of earth to keep it from opening onto Gloucester Harbor, which would put the Atlantic on both of its ends.

As early as 1638 there were discussions of cutting through these last few yards. This would give the people of Annisquam, and anyone else living on the Ipswich Bay side of Cape Ann, a water route to Gloucester Harbor or Boston without taking the treacherous voyage around the Cape that the tragic Thacher and Avery families had attempted on the *Watch and Wait.* At the time, seagoing vessels were small enough to navigate the Annisquam. But Gloucester Harbor was only a little fishing port, and according to the colonial Massachusetts government, it did not justify the expense.

Frustrated by the lack of government support, a town meeting in April 1643 voted to turn over the project to a private commercial venture taken on by the town minister, the same Reverend Richard Blynman from Plymouth. He had the Cut dug and shored up with rocks that joined together at the bottom and tapered out as they rose so that the stone passageway was just wide enough for one small shallop, which was little more than a dinghy with a sail. A bridge was built over the opening, still known today as the "Cut," that was anchored with a swivel on one side and could swing open or shut. A storm in 1704 caused high tide to fill in the Cut with so much sand and gravel that it was no longer navigable. This happened again in 1723. But the town kept clearing it out. In 1830 the town decided to abandon the waterway, and it was deliberately filled in but was reopened in 1868.

There was a significance to the Cut beyond facilitating shallops from the other side of the peninsula. Once those few yards of dirt were dug, Gloucester became an island, and its inhabitants have thought with the insularity and self-sufficiency of an island people ever since. Until the interstate highway system was built in the 1960s, the rest of the world remained far away, and many Gloucester people liked it that way. Even though Boston was visible from Gloucester Harbor as tiny bumps on the horizon across the sea on a clear day, to the people of Gloucester it was a distant city. You had to go over the Cut bridge. Today the little bridge goes up and down instead of twisting to the side, and it carries two lanes of cars over a canal widened in the early twentieth century, but it still represents a barrier that must be passed between Gloucester and the world. Many Gloucester people spent their whole lives without ever leaving their island, except that they might travel hundreds of miles out to sea. Others crossed the Cut into Massachusetts on occasion, but leaving the island was always significant. Not without irony, twentieth-century poet Charles Olson wrote that he did not often visit his father "living over the Cut as he did."

The island of Gloucester was a reality on the map and in the minds of the residents. Rapidly it became a socioeconomic reality as well. Gloucester was the most self-sufficient fishery in New England. The first boat was built in Gloucester in 1643, the same year the Cut was dug. Gloucester workers built their own boats from the lumber of local trees that Gloucester workers milled themselves, and they repaired the vessels themselves. They had a foundry that made anchors and iron fittings, they had sailmakers and rope makers, and they even produced their own oilskins, the bright orange waterproofing that fishermen wore. Gloucester developed separately from the world around it. Sefatia Romeo, asked why Gloucester Sicilians kept their traditions more than Sicilians elsewhere in America, said, "We are different in Gloucester. Our life doesn't step over that bridge."

Gloucester society was different because it was a society completely centered on fishing. According to legend, in those early Puritan times, one Sunday in church the minister reminded the brethren that they had come to this new land to worship the Lord, whereupon one of the brethren added, "and to catch fish!" Fishermen share work, share risks, and share the take. It is a commercial activity that promotes egalitarianism.

Today the Annisquam is a kind of border with its shifting, sandy tidal banks where soft-shelled clams bury themselves in stiff vertical poses. Officially the opposite bank is still Gloucester, but it is a rural, wooded, agricultural Gloucester. Manchester-by-the-Sea and Essex, the two closest communities on the other side of the Annisquam, are more typical New England towns, with an agricultural feel, a sense of space and country life, and country gentry, of WASPdom. They have their maritime traditions—many of the Gloucester schooners were built in the marshy Essex waterfront—but they are coastal towns that look in at the land. Gloucester always looks out to sea.

By the mid-seventeenth century, 10 percent of the population of Ipswich, later to become Essex, owned almost half the wealth of the town. The wealthiest merchants of Ipswich had amassed huge fortunes. Gloucester was a different kind of society. It had almost no wealthy or poor people; the fishing economy made for only small differences in income. By the end of the century in Gloucester, the wealthiest 10 percent of the population still had only one-third of the wealth and were worth only about a quarter of what the richest families of Ipswich were worth.

While Ipswich granted huge tracts of land to individuals, Gloucester kept the large parcels as common town land. On the other hand, Gloucester, which always depended on new-comers to maintain its population, granted new arrivals far more generous parcels of land than did Ipswich, so that new-comers very quickly had the same size holdings, and therefore the same standing, as long-time residents.

These early beginnings embedded an egalitarian spirit in Gloucester that has never left, not even when truly wealthy people moved to town. Even today people frequently refer to "the blue-collar spirit of Gloucester."

By the end of the seventeenth century Gloucester became a peaceful, egalitarian society. The crime most often brought to Gloucester courts was violation of the Sabbath. The defendant was rarely from Gloucester. Usually it was a Boston vessel that decided to set sail on a Sunday morning, a crime that Glouces-ter always prosecuted. The rest of the court cases were not that different from today's Gloucester police blotter: a few cases of drunkenness, a few cases of domestic violence, a dispute at a rowdy tavern. Young men would go to the taverns before church services and "some doe very much indispose themselves for the worship and service of God."

By 1700, Gloucester had already become Gloucester.

chapter Four

SCOONING

. . .

Also pray for those who were in ships, and
Ended their voyage on the sand, in the sea's lips
Or in the dark throat which will not reject them
Or wherever cannot reach them the sounds of the sea bell's
Perpetual angelus.

—T. S. ELIOT,
"THE DRY SALVAGES,"* NO. THREE OF *FOUR QUARTETS*, 1941

A DECADE AFTER JOHN CABOT'S 1497 VOYAGE OPENED UP
Newfoundland fishing to Europeans, fishermen had discovered
the Grand Banks and concentrated their fishing there. But in
Gloucester the inshore fishing was so good that it took them
two centuries before they went out to the banks. This is characteristic of Gloucester fishermen. They have always tena-
ciously stuck to a thing as long as it is working. In almost four
centuries, commercial fishing in Gloucester has paradoxically
remained dominant while at the same time, old-fashioned.

In 1645 the first boats went to the Grand Banks, though
Gloucester fishermen continued to pursue inshore fishing, especially for mackerel. The earlier boats were either ketches—
two-masted vessels with fore-and-aft rigging—or sloops, boats

*"The Dry Salvages" are an exposed rocky reef at sea off the end of Cape Ann.

that were also fore-and-aft rigged but usually single-masted with a mainsail and a jib, sometimes with topsails, but far less sail than a schooner. Fore-and-aft rigging, as opposed to square rigging, means the sails ran from bow to stern instead of perpendicularly. Both had open decks or no decks at all, no place to take shelter in rain or sleet. Then came pinkies—two-masted vessels with no jibs, which are the triangular sails over the bow. Pinkies were used in Gloucester well into the nineteenth century, the last one built in 1844. They had a cabin and a six-inch edge to the deck that was supposed to keep fishermen from getting washed overboard.

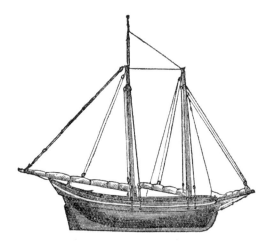

Pinkey, 1810

And at least they had two bunks and a fireplace for cooking. For long voyages the men carried hard sea biscuits, soaked for softening before eating, which were the first American crackers, and fat-back pork, which they would cook with fish—the origin of chowder. They avoided eating cod because it brought

Wingeaersheek Beach

such high prices, preferring halibut, which was little valued until the nineteenth century and kept well when smoked.

Into the twentieth century, Gloucester and Cape Cod fishermen were said to be particularly well-fed, even though before refrigeration theirs was a cuisine completely devoid of any perishables other than fresh fish. This recipe, which—poetic license aside—follows the tradition of fishing boats at sea, was published in the *Boston Evening Post*, September 23, 1751.

Directions for making a CHOUDER

First lay some Onions to keep the Pork from burning,
Because in Chouder there can be no turning;
Then lay some Pork in Slices very thin,
Thus you in Chouder always must begin.
Next lay some Fish cut crossways very nice
Then season well with Pepper, Salt and Spice;
Parsley, Sweet-Marjoram, Savory and Thyme
Then Biscuit next which must be soak'd some Time.

Thus your Foundation laid, you will be able
To raise a Chouder, high as Tower of Babel:
For by repeating o're the Same again,
You may make Chouder for a thousand Men.
Last Bottle of Claret, with Water eno' to smother 'em,
You'l have a Mess which some call Omnium gather 'em.

In the nineteenth century, with the advent of canning, milk-based chowder became available to fishermen. This mid-twentieth-century recipe from the Gloucester seafood company, Gorton-Pew, is more commonplace today.

> Place the contents of one can of Gorton's Down East Clam Chowder in a saucepan, adding an equal quantity of milk. Let it barely come to a boil and serve.

THE FISHERMEN HAND-LINED, often two lines per man, paying out and reeling in on their thumbs more than three hundred feet of line over and over again, standing all day on the edge of the sloop deck in most kinds of weather. In Gloucester the townspeople said they could always recognize a hand-liner from his stooped-over, round-shouldered posture.

But as Gloucester fishermen fished farther offshore, finally going to Georges Bank and even the Grand Banks, with characteristic reluctance, they recognized that they needed a change in technology: a faster, long-distance fishing boat. In 1713, a date now remembered as one of the most important in Gloucester history, Andrew Robinson, a fishing captain, launched a fishing schooner in Gloucester. According to Gloucester legend, the first schooner was thus invented. But his two-masted schooner was actually similar to the very large ketches used by the Dutch a century or two earlier. The only differences were that on a schooner, the aft mast is taller and the vessel has topsails. Robinson's schooner may have been the first boat to be called a schooner, though, and was the first of the famous Gloucester fishing schooners that would dominate the fisheries of New England and maritime Canada well into the twentieth century.

Schooner is an odd word of uncertain origin. According to Gloucester legend, someone watching Robinson's new boat sailing around Gloucester exclaimed, "She scoons!"—or even "There she scoons," in the saltier version of the story—*scoon* being a Scottish word meaning "to skim lightly across the water." This improbable legend stands for lack of any better one, though it leaves etymologists scrambling to explain the *h* in schooner, which seems to indicate German or Dutch origins.

The schooner was built for speed. The ratio that creates speed is the greatest possible amount of square feet of sail, driving the fewest possible cubic feet of hull space through the water. Previous fishing boats had beamy hulls, a lot of width on which to place a deck with plentiful space for cleaning and curing the catch. But the new, swift schooners had narrow decks and clouds of billowing sails aloft. They were fast, graceful, and unstable. Fishermen started putting even more canvas aloft to sail even faster, racing for records back from Georges Bank until a jumble of rocks, trees, and distant hills showed in pur-

An 1876 advertisement showing a schooner

plish lumps near the horizon and the two towers of Thacher Island were at last seen off the starboard.

Schooners raced each other. From 1886 to 1907, a series of Canadian–U.S. schooner races took place that essentially pitted the Banks fishermen of Gloucester against the Banks fishermen of Lunenburg, Nova Scotia. But the speed of schooners was not developed for racing; it was meant to get fish to market quickly. As speed became a fascination, a market for fresh fish from the Banks developed.

There was not much yachting in New England until the mid-nineteenth century. Even in the 1850s, when yachtsman began racing in what was to become the America's Cup it was of little significance to Gloucester fishermen—simply rich people playing. It was inevitable that the rich would start playing with schooners, for they were not only swift, they were also beautiful. For almost a century, the wealthy of New England frolicked on cleaned-up versions of the same boats the fishermen used. In 1920, when the America's Cup was called off because of foul weather, the publisher of the *Halifax Herald and Mail* awakened fishermen's pride by claiming that yachtsmen did not know how to handle schooners and fishermen should race them. The *Gloucester Daily Times* took up the challenge and Gorton's, a leading Gloucester seafood company, sponsored a schooner that twice beat the Nova Scotians. Then after a season fishing the Grand Banks, the *Bluenose*, a large schooner forty-nine meters long and only eight abeam, sitting only five meters in the water with a 36-meter-high foremast and more than a thousand square meters of sail, defeated the Gloucester *Elsiea*. During the next seventeen years the *Bluenose* remained undefeated. The Canadians have put the image of the *Bluenose* on postage stamps, the Canadian dime, matchbooks, and seemingly most spare surfaces in Nova Scotia.

But the defeat was not important to Gloucester. Fishermen and their families had other concerns. Since the introduction of schooners, the world's most dangerous trade had become even more dangerous. The vessels' narrow hulls and voluminous canvas blew over easily in a gale. The early schooners had no scuppers—holes in the side to let seawater run off the deck. In an effort to make the long trip pay as much as possible, the schooners were so overloaded that they would sit deep in the water, and if a storm came up, it would take only a few large waves to sink them.

The schooners, for all their speed, made three-month trips to the Grand Banks. Gloucester schooners did not fish the closer Georges Bank until the 1830s, when they started find-

Rebuilt Bluenose *visiting Gloucester harbor in 2005*

ing cod stocks there more plentiful than those in the Grand
Banks. Was this the first early sign of overfishing the Canadian
banks? Not enough records were kept to know, but from then
until the late twentieth century, when depleted stocks and fishery management drove them off, Georges Bank was the leading grounds of Gloucester fishermen.

Georges Bank is a treacherous shoal, 150 miles east of
Gloucester. The tides run extremely strong and can drive a
schooner aground. It was discovered that the best fishing there
was during the even more treacherous winter months. Georges
Bank was more crowded than the larger Grand Banks, as the
schooners all tried to weigh anchor and fish in the choicest
spots. When the seas were high in open ocean, a skillful
schooner captain would head into the wave and climb to the
whitecap and slowly back down the other side without great
danger. But on Georges Bank, because the sea was shallow,
waves were not long, rolling swells but short, high chops that
would smack into the schooner like a flurry of punches, some-

times for hours. Having other schooners anchored close by meant a great risk of crashing into each other in a storm. As the vessels were smacked by foaming mast-high waves driven by gale-force winds, the strong Georges Bank tide would run in the same direction as the anchor, and even two anchors could lose grip so that the schooner would drift into the shoal or the other vessels. The result of such a collision between two schooners was that both ships and all hands were lost.

Several times, more than a hundred Gloucester fishermen were lost on Georges Bank in a single storm. Twice, Gloucester lost more than a third of its fleet in one night.

Family and friends back in Gloucester had little information other than which schooners returned. A report on the front page of the March 19, 1862, *Gloucester Telegraph* shows the kind of guesswork by which Gloucester counted its dead. The lead story, upstaging an account of General McClellan and the Army of the Potomac, begins:

> The gale of the 24th and 25th of February last has proved most disastrous for our fishing fleet and from present advice the loss of life will be very great, largely exceeding that of any previous season.

The article goes on to speculate that, in addition to the known sinkings—vessels that were seen crashing into others during the gale on Georges Bank—there were still "thirteen vessels that have not been heard from since the storm, and we fear never will be." The newspaper argued that other vessels that were caught in the gale have "been in, gone out and arrived a second time" since then. So it was thought unlikely that the missing vessels were still out fishing. The article went on to estimate that between 115 and 120 fishermen were missing.

A large portion of them leave families and it has been estimated that as many as five hundred children will be made orphans by this disaster; and great distress will be caused by the loss of so many upon whom others were dependent for support, while a gloom has been cast over our community.

Gloucester had long been a community of widows and orphans. Occasionally a missing schooner would miraculously appear rounding Eastern Point. On shore, eyes were fixed on the horizon. Sometimes a vessel would be sighted coming home only to sink while wives and children were watching. But in this case the prediction was sadly accurate. One hundred and twenty Gloucester fishermen were lost on Georges Bank from the gale of February 24, 1862. The 1862 *Gloucester Telegraph* wrote:

This winter-fishing upon Georges Bank becomes more disastrous year after year, as the numbers increase who engage in it. Yet men will follow it, and owners of property will risk their vessels, knowing with a certainty that all cannot escape—

These horrors have become part of the collective memory of Gloucester even though most of the current families had not even arrived at the time they happened. In 1766 nineteen vessels were lost in a storm while sailing to the Grand Banks. In the 1870s, a time when enormous amounts of sail were rigged aloft to make ever speedier schooners, losses became even more staggering. In 1871 twenty schooners and 140 men were lost. In 1873 nine vessels and 128 men were lost on Georges Bank and North Bay in a summer gale. In February 1877 thirteen vessels and 143 fishermen were lost in a gale on Georges Bank. In 1878 twenty-nine vessels and 249 fishermen were lost.

Predicting the weather was less than a science. In 1643 an Italian scientist, Evangelista Torricelli, invented the barometer. If it rose steadily it usually meant fair weather was coming, and if it fell rapidly this indicated an approaching storm. Fishermen could also stare up at the sky and interpret the significance of clouds and patches of blue, shifts in the wind, interpreting patterns they had observed during years working the same corner of the ocean. None of this was dependable enough to stake your life on, but that is what they had to do.

Even the notion that home was safe, that there was no more danger after reaching the deep and stately protected harbor in the lee of the wind that had so impressed Champlain, became a shattered illusion in 1839.

Saturday, December 15, 1839, was an unseasonably warm mid-December day and captains all over the northeast decided to go to sea the following day. Sunday morning, hours after they set sail, the wind shifted fiercely to a southeaster. Quickly the many vessels off Cape Ann put into the sheltered harbor of Gloucester to sit out what was quickly becoming a gale. The crews struggled helplessly as the gale winds swept across the crowded harbor, crashing ships into other ships or splintering them on the rocks along the harbor's western shore, as the townspeople watched on, unable to help. The destruction continued through the day. By nightfall twenty ships were destroyed but by morning another thirty had lost their masts or had them chopped off to keep the hull from being blown over. Vessel and crew could be lost without even going to sea.

WHY DID THEY ENDURE the risks of a winter fishery? Gloucester fishermen were not a desperate, backward people. According to ichthyologist George Browne Goode, in his seven-volume 1880s treatise on fisheries, Gloucester fishermen, unlike those of Maine, were educated and read periodicals and even Shakespeare and Dickens, and they lived in sturdy, comfortable houses—many of which are still in use. They risked winter fishing for the promise of year-round employment and a prosperous economy.

Improved technology was usually directed at increasing catches and not at improving safety. Fish were not seen as a finite resource until the twentieth century. It was widely believed that fish, especially cod, because they laid millions of eggs, were indestructible. A basic concept of nature—that animals whose young have a poor survival rate bear large numbers of offspring—was not understood. A codfish might lay a million eggs and that million could be scooped up and eaten, every last one, in a single pass of a humpback whale. Almost everything that swims, along with high seas and foul weather, is the natural enemy of fish eggs and young fish. Even the preeminent Darwinist, Thomas Huxley, did not understand this, and as late as 1883 was still insisting that "anything like permanent exhaustion" of a fish stock from overharvesting was impossible.

And yet the people of Gloucester did understand the concept of a finite natural resource. They treated trees that way, recognizing that Cape Ann could produce barely enough lumber to build the Gloucester fishing fleet. They would have found it unthinkable to have given up their island's self-sufficiency by bringing in lumber from elsewhere, and so from the first years there, they placed restrictions on the private use of trees on Cape Ann's public lands. The town licensed lumber mills, regulated prices, and restricted out-of-town sales. In 1702 the people of Gloucester voted at a town meeting to further restrict lumber sales outside of Gloucester. Shipbuilders now had to sign a statement guaranteeing that a vessel built of local trees would be sold only to someone from Gloucester. Fines and other penalties were enforced against anyone who excessively harvested wood, and rarely did townspeople violate these regulations.

But, from the beginning they saw fish as limitless. "Though there be fish in the Sea, foules in the ayre, and Beasts in the woods, their bounds are so large, and we so weake and ignorant, we cannot much trouble them," John Smith wrote of New England: The only obstacles seen to catching more were the limitations of technology. In 1630, Francis Higginson wrote:

> Of our fish our fishers take many hundreds together, which I have seen lying on the shore to my admiration;

yea their nets ordinarily take more than they are able
to hale to land, and for want of boats and men they
are constrained to let many go after they have taken
them. . . .

Fishermen were always looking for a way to take more fish,
and the next idea—long-lining—made fishing even more dangerous to fish stocks. The European idea of fishing from small
dories did not interest New Englanders until they embraced
long-line fishing in the nineteenth century. Long-lining was
done with a long, heavy line, known as a ground line, with
shorter, lighter lines fastened to it every three to four feet.
These shorter lines had baited hooks.

The line was called a trawl, a bulltow, or a setline, and it
was not a new idea. There is a record of long-lining off of Iceland in 1482, and it may not have been a new technique even
then. But in 1815, after years of war—the French Revolution
and the Napoleonic wars—the French government decided to
rebuild their fishery, subsidizing a fleet of long-liners. This
program greatly upset other countries. The British suspected
that the French were rebuilding their fisheries as a clandestine
way to rebuild their navy. It was commonly felt that fisheries
were a good training ground for naval forces, and John Adams
had argued vigorously for the new American Republic to support the New England fisheries for that reason, claiming that
supporting fisheries was more cost-effective than maintaining
a large standing navy. Gloucester fishermen had proved very
useful in naval engagements of the American Revolution. At a
time when the entire population of Cape Ann was estimated to

be about 5,300 people, 1,565 Gloucester residents fought in the Revolutionary War; in all the wars since, Gloucester has always contributed disproportionately, especially to the Navy.

The growth of the French long-line fleet was protested not only by other nations that suspected France of renewed militarism but also by fishing nations that protested the French government subsidy as unfair competition. The Scandinavians insisted that long-lining was an undemocratic way of fishing because it required a wealthy company to buy the large quantity of bait needed. But absolutely no one protested that this fishing technique, which greatly increased the catch, would lead to a dangerous reduction in the fish population, known today as the stock.

When long-lining was introduced to New England, the only objection was that with so many long-lines working the same fishing ground, they would constantly be fouling each other's lines. The solution was to launch two-man rowboats—dories—about sixteen feet long. The schooners would head to the banks with six dories, stacks of three each on both sides of the ship, with several thousand herring to bait the hooks.

The dory, a solidly reenforced rowboat built to be tossed in a rough sea, was loaded with long spruce oars and wooden tubs in which the trawl was coiled, and then it was lowered to the sea. The dorymen, bulky in their yellow oilskins, would jump over the side of the schooner into their dories. They would row out, all six dories trying to head in parallel paths on the windward side of the ship so that when the time came to return, they would have the wind at their backs, blowing them toward the ship. Once out far enough, the man in the bow would drop the anchor of the trawl, and the tarred cotton line about a quarter inch thick would pay out of the first tub until the anchor landed on the bottom. Halibut and cod both feed on the bottom. Once the anchor was set, the doryman would let out a caulked barrel with a flag on top as a buoy. The buoy attached to the anchor marked the end of the trawl, which was coiled into four tubs in fifty-fathom lengths called skates. The skates were payed out with baited hooks every fifteen feet. The doryman would knot each skate to the next, with six skates in each tub. After the last skate of the last tub, there was another anchor and another buoy to mark it, some mile and a half from the first one.

A doryman accomplished all this, laying almost five hundred baited hooks on the bottom while his dorymate maneuvered in the high, short chop of Georges Bank. The little boat had to be constantly rowed about so that it did not get hit broadside by a wave. Fishermen usually set their trawl in late afternoon. Then they returned to the mother ship, exhausted, and the next day took the dory out in the high sea again to haul back the trawl, which now had large, heavyset Georges Bank codfish or two- and three-foot halibuts twitching from the hooks. While the one mate managed the oars and the trawl, the other was at the bow with a hooked pole, wrestling the fish onto the boat. Codfish were not uncommonly four or even five feet long, and halibut ran from two hundred to four hundred pounds and had to be hauled onto the little rowboat while it was bobbing from trough to trough, between the white peaks of a high sea. Incredibly, the Portuguese fished the Grand Banks well into the twentieth century with only one man in a dory.

In Gloucester, halibut dorymen were said to be the largest, toughest men. But they were all tough. In 1876 a Danish-born Gloucester doryman named Alfred Johnson was dared to sail his dory across the Atlantic. Fifty-eight days after leaving Gloucester, he arrived at Abercastle, Wales, which was the first record of a one-man North Atlantic crossing.

There were even more ways to die in a dory than on a schooner. The little open-decked rowboat could capsize by hitting a wave broadside. As the dory filled with fish it sat ever deeper in the water, which made it both more difficult to maneuver quickly and less likely to ride out a high wave. Sometimes the weight of the fish became too much; with one fish too many, the entire dory, two dorymen, fish, and all would suddenly sink straight down, swallowed up in an invisible "dark throat," as T. S. Eliot called it, remembering stories from his childhood summers in Gloucester. The dorymen were constantly caught between the urge to catch as many fish as possible and the fear that the next fish would sink them.

The fishermen who worked in these dangerous conditions were exhausted from sleeping only a few hours a night. When they got their catch back to the mother ship, they had to clean, perfectly split, and salt the cod or ice down the halibut before going to sleep. Their fatigue contributed to a large number of fatal accidents, including falls overboard.

Often, the dorymen would be out fishing in the fog or the darkening winter afternoon with their boat fully loaded with fish and then, looking around, realize that they had lost sight of the mother ship. They could sit it out in the sea in their sixteen-foot rowboat until the next daybreak or until the fog

A dory at sea in early 1900s
(PHOTO BY HERMAN W. SPOONER, COURTESY OF THE CAPE ANN
HISTORICAL ASSOCIATION, GLOUCESTER, MASSACHUSETTS)

lifted, they could row around searching for their ship, or they could try to row hundreds of miles to safety. Any of these three choices could result in the dorymen's freezing or starving before they were ever found. A long-standing Nova Scotia record in the nineteenth century was held by a doryman who was picked up alive after being lost in the fog for sixteen hours.

According to New England fishing lore, the two mates who work a single dory are closer than the closest of brothers. They depend on each other for survival. Stories abound of one dorymate risking his life, even giving his life, to save the other. In the early twentieth century, most of Gloucester knew the story of John Rose, whose larger dorymate gave him his sweater to stay warm when they were lost one night at sea.

Rose kept asking his mate if he needed it back and his mate
kept insisting he was warm enough, until the moment in the
dark when Rose called out to him and the mate didn't answer
because he had frozen to death. A Gloucester story.

THE MOST FAMOUS Gloucester story of all time is about a
halibut doryman named Howard Blackburn. On January 23,
1883, Blackburn, a Gloucester fisherman who had emigrated
from Nova Scotia, and his dorymate, Tom Welch, lowered their
dory from the *Grace L. Fears* and began to haul in the trawl
they had set only hours before. Blackburn and Welch had
shipped out on Gloucester's best halibuter. The year before,
Captain Nathaniel Greenlief had taken the *Fears* on a five-week
trip landing a record fifty tons, the most profitable halibut trip
in Gloucester history. This meant that the *Fears* could hand-
pick the best halibut dorymen available on the New England
and Nova Scotia coasts. When the schooner put in to Nova
Scotia to let off a sick man, Greenlief decided to stay ashore
and turned the trip over to the first mate and cook, Alex Griffen. Short a doryman, Griffen hired on a man he had known
and trusted from Gloucester, Howard Blackburn.

Griffen took the schooner to Burgeo Bank, sixty miles
south of Newfoundland, a small bank before the Grand Banks.
They set the trawl, Tom Welch probably noting that the new
man who was assigned as his mate rowed in powerful strokes,
his large hands curled around the oars. Only twenty-three
years old, at six feet two, with more than two hundred pounds

of bone and muscle, he looked like a halibuter. After they set the trawl they returned to the schooner, but only two hours later, Griffen, feeling a wind begin to blow, ordered the dorymen back to their boats. With so little time, the catch would be small, but Griffen was a cautious skipper and he did not want to risk his crew on a stormy sea. As the men rowed back to their trawls, the first snowflakes were flittering above the sea.

Welch and Blackburn hauled in their trawl. The other dories were a little ahead of them. Perhaps these two had more fish on their lines. Blackburn stood midboat with a club, stunning and unhooking the fish, like huge white platters. Just as their work was done, the storm came up from the opposite direction, meaning that they were now in the lee of the ship and had to row into the wind to get back to the *Grace L. Fears*. It is easy to become disoriented in a storm at sea, and staring through the pelting snow and only able to see as far as the bow of their small dory while Blackburn pulled hard at the oars, the two lost all sense of where their schooner lay. Had they gone past and not seen her? It was possible. The storm howled and they could neither hear nor see. Bailing and rowing, Blackburn lost his rowing mittens over the side and his fingers began to turn white. Realizing that his hands were freezing he wrapped his fingers around the oars and kept them there until they were frozen in rowing position.

As night fell the snow stopped, and to their surprise they saw that they had not passed their schooner, whose signal torch was visible in the rigging far ahead of them. But when they tried to row toward her, the wind was so strong and the sea so high that they made no progress. The waves were crashing in and they desperately tried to keep afloat in the freezing sea, bailing out the seawater with a buoy keg. Every now and then while tossed on the crest of a wave, they glimpsed the signal torch faintly in the distance. But by morning they saw nothing but endless sea and the shine of the ice varnishing the sides of the dory.

They took turns rowing and bailing, but after a time the younger Welch was exhausted. By the second night he had frozen to death. Blackburn rowed by himself for one hundred miles with the frozen body of his dorymate, and somehow he reached Newfoundland.

He lost all of his fingers and most of his toes, and it would have been understandable if this had been all he ever wanted to see of the ocean. But Blackburn had a sloop specially designed for his disability and went on to set a thirty-nine-day, one-man Gloucester-to-Lisbon record. He continued on various other nautical adventures as well, including rowing the Florida coast with oars that were strapped to his wrists. He also captained a ship around Cape Horn and tried, unsuccessfully, to get in on the Klondike Gold Rush. Later in life, he owned a bar on Main Street in Gloucester from which he dispensed money to anyone in need who stopped by. It has been estimated that the few dollars he handed out regularly from a drawer eventually totaled $50,000.

After Theresa, his wife of fifty years, died, Blackburn at age seventy-one, white haired but still large and robust, started to weaken physically. He began to feel the effect of years of untreated wounds, including a hernia caused by trying to lift the frozen body of Welch out of the dory in Newfoundland. In

1932 he died at seventy-three and was buried in West Glouces-ter, in a cemetery known for being a resting place for the fortu-nate fishermen who got to die on dry land.

More than seventy years after his death, there were still a few people in Gloucester who remembered Blackburn. Joe San-tapaola, a nonagenarian retired fisherman born in 1914, re-called seeing the way Blackburn would scoop money off the counter with a hand with no fingers and how Blackburn would stand in the doorway as school let out and pass out candy to the children as they ran by. The children wanted the candy, but they were afraid of his fingerless hand.

Today Blackburn's bar is a popular restaurant with Black-burn's fingerless portrait displayed in the bar. A tavern was named after him, as was Gloucester's first industrial park, and a traffic rotary. Blackburn's is still remembered as the greatest of Gloucester stories.

DEATH AND SURVIVAL AT SEA, Welch and Blackburn, re-mained the Gloucester way of life. From 1830 to 1900, the golden age of the Gloucester schooner, 3,800 Gloucester fish-ermen and 670 schooners were lost at sea. And the toll contin-ued throughout the twentieth century. Families would keep their eyes fixed on the outer harbor, straining for the sight of gray sails rounding Eastern Point. By the time a schooner passed the little green hump called Ten Pound Island, they could see which one had made it back and whether it was flying a flag at half-mast, which meant that not all the crew were coming home. The year 1912 was typical. That year, ten ships and forty-five crew members were lost, which is neither one of the better years nor one of the worst. Eleven of the dead were lost in shipwrecks, seven men died in dories that capsized or sunk, nine vanished in dories that were lost, seven were knocked overboard, five died on their ships, two drowned at the wharf, two were killed in onboard accidents, one died in a hospital from an injury resulting from an onboard accident, and one man was actually struck by lightning while on deck and knocked overboard and drowned.

In 1923, for the three-hundredth anniversary of Gloucester, a boulevard was created along the beginning of the western side of the harbor, just past the Cut. Sculptor Leonard Craske was commissioned to build a memorial to Gloucester's unending losses. *The Man at the Wheel* depicts a fisherman in oilskins standing at the wheel—some argued that he is on the wrong side of the wheel, out of view of the compass. But that man, known as "the Old Salt," is how Gloucester defines itself.

Incomplete records going back to 1623 suggest that as many as ten thousand Gloucester fishermen may have been lost at sea. "And yet men follow it," as the 1862 *Gloucester Telegraph* marveled. What drove them to keep doing this dangerous work? Not greed, because not one of these men risking their lives at sea ever amassed a fortune. Rather, they were driven by the same forces that drove artisans and craftsmen like cabinetmakers and silversmiths: a sense of a unique place in society, a pride in bearing the title "fisherman," a desire to be a great fisherman.

A fisherman is not a salaried worker. Even on early European vessels fishermen worked for a share of the profits from the catch. In Gloucester the catch was divided fifty–fifty between ship and crew. Sometimes a trip would be a "broker." Other times there was enough profit for everyone to take home a big paycheck. When Howard Blackburn finally got home without fingers or toes, the five days he was lost at sea and not fishing were deducted from his pay. He brought home $85, which at the time still would have been considered far short of a big nest egg, but a substantial paycheck for most laborers on land.

A fisherman is proud to work for himself and earn his living on the sea, which makes him a little better than everyone sadly bound to the land. Men who face this much danger, similar to combat soldiers and police, have a closeness and camaraderie, feeling they belong to a special brotherhood—and by extension so do their families; their town is itself a special close-knit fraternity.

Then, too, there is the inexplicable joy of yanking a fish up from the water. A young man who had worked in the far less perilous summer mackerel fishery read of the winter Georges Bank fishery and went to Gloucester in search of adventure. After surviving the fatal February 24, 1862, Georges Bank gale, he wrote of his experience:

> The cold, to one of my constitution, was intense, and pierced into the very marrow of my bones, although thickly clothed. But this deep sea fishing was so exciting that I stood at the rail sometimes a full hour, without changing my position, pulling in the big codfish and occasionally a halibut. It was a moment of extreme gratifi-

cation when I hauled in my first fish of the latter species, and saw him floating alongside with the hook securely fastened in his mouth.

Gloucester was populated by simple men with modest incomes and without affectations, but who walked with an aristocrat's belief that they were a special breed—a breed that was never curbed by fear and knew things that other men missed. What was the effect of constantly losing a few or a few hundred, tragedy upon tragedy, as family, friends, and neighbors simply vanished into unmarked icy graves? It created a deep sense of community, of commitment to the fishery. The Cape Ann granite beneath *The Man at the Wheel* bears these words: "They that go down to the sea in ships." The rest of the psalm was so well known in Gloucester that there was no need to write it out.

> They that go down to the sea in ships,
>   that do business in great waters;
> These see the works of the Lord,
>   and his wonders in the deep.
>   —Psalms 107:23–24

chapter Five

THE REPLACEMENTS

. . .

The brilliant Portuguese owners,
they do. They pour the money back
Into engines, into their ships,
whole families do, put it back

—CHARLES OLSON,
"LETTER 6," 1950

UNLIKE TREES, MORE LIKE FISH, IN GLOUCESTER PEOPLE
were always seen as a renewable resource. The town had always found new immigrants to replace its losses.

Just as in the very early years of settlement, those who left
in disputes were always quickly replaced by new arrivals who
had heard that there might be opportunities in the Cape Ann
fishery. As hundreds disappeared in storms, they too were replaced by fishermen from other Atlantic fisheries. That is how
Gloucester evolved into an Irish, Scandinavian, Portuguese, Sicilian town. The Puritan village became a place of Catholic ritual, of linguica sausages, pasta, and cannolis. By 1900, 40
percent of the 26,000 people in Gloucester were foreign-born
and the percentage continued to rise.

In the 1820s, Newfoundland and Nova Scotia fishermen
started coming, which meant fishermen of Irish background,
fishermen like Howard Blackburn, though Blackburn's family

were actually Tory Americans who fled to Nova Scotia after the Revolution. By 1900, the majority of skippers were from Newfoundland and Nova Scotia. But there were also Scandinavians. When in the early twentieth century the Dutch craved Cape Ann–built schooners, they would pay a crew to sail one to Holland. Scandinavian fishermen would take the offer so that they could visit home again. But the Dutch did not pay the return passage. It mattered little to Gloucester. There were others to take their places—more Scandinavians, more Irish from Newfoundland and Nova Scotia, looking for better fishing grounds and a better life.

Gloucester offered work not only for the men fishing in Georges Bank schooners but also for the women, using their seamstress skills to make oilskins or working in the nearby

Fish flakes in Rocky Neck, circa 1900
(COURTESY OF BODIN HISTORIC PHOTO)

mills of Lowell, Peabody, and Lawrence. Women who were not skilled with needles could find work along the Gloucester waterfront, maintaining the fish flakes—the outdoor wooden racks for drying split and salted fish that sprawled along the edge of the harbor from the center of town to East Gloucester. Carpenters and laborers could work in the shipyards of the nearby wetlands of Essex.

Until the twenty-first century, Gloucester remained a community almost entirely dedicated to its fishery. In addition to work as fishermen, there was always work on the docks for "lumpers," who unloaded boats, operated forklifts, chopped ice, and iced down catches. Policemen and firemen supplemented their income as part-time lumpers. Children earned a quarter to pack a net into its box. Almost everyone in town worked at least part-time for the fishing industry.

Fishing even provided work for artists. The first ones were native born such as Fitz Henry Lane, the son of a Gloucester sailmaker. He was born Nathaniel Rogers Lane in 1804 and later changed his name to Fitz Hugh and then Fitz Henry Lane. This Lane of the shifting names was not the first painter to work here, because maritime painting was already a popular genre. But he was the first Gloucester painter of note, the one who put Gloucester on the map of American art. Ever since, painters, as well as fishermen and dockworkers have come to Gloucester looking for work.

Originally Lane had been known for his accurate depictions of Gloucester vessels. Although handicapped, possibly from childhood polio, he would get crews to hoist him aloft to study rigging. It was said that you could rig a schooner from a Lane painting. That was the influence of his having grown up in Gloucester. After studying in Boston, he became more interested in light than maritime accuracy. He was known as one of the foremost Luminists—nineteenth-century painters of romantic landscapes with hidden brushstrokes and a lyrical use of light, who were predecessors of the Impressionists. Gloucester has a reputation for extraordinary light. This is partly because Lane's famous seascapes created a lasting image of Gloucester and its harbor. But it is also because Gloucester is built on a series of inlets and peninsulas so that there is water everywhere, creating reflected light softened by humidity, periodically turning the most mundane of industrial sights into scenes of heartbreaking beauty. Lane could look out the window in the granite house he designed in Gloucester Harbor and view raspberry mornings and lilac sunsets. Many of his most celebrated works were painted from a third-story window.

Lane was a popular figure in Gloucester and as his reputation grew, other artists began to arrive, especially other Luminists, such as Francis Augustus Silva, one of the first of many New York painters to come to Gloucester in a steady flow that continued long after Lane died in 1864.

AMONG THE OTHER non-fishermen drawn to the Gloucester fisheries were garment makers who produced clothes for fishermen. Many of them were Jewish immigrants who started arriving in Gloucester in the late nineteenth century. Like the Irish before them and the Sicilians after, they were met with

An 1876 advertisement

some suspicion. Older Jews still remember accusations of being "Christ-killers." Harold Bell, born in Gloucester in 1914, only remembered being excluded from clubs he did not want to join. Another Gloucester Jew, Ida Marks, said, "We got along. They called the Italian kids 'Wops' and we were the 'Hebes' and the Portuguese were something else. It was just the way they described other people. Certainly it would have been nicer if they hadn't."

Bell was the son of Morris and Molly, Jewish immigrants who had owned a well-known shop in the garment trade of New York's Lower East Side. Morris knew Jewish people from his native Bialystok who had settled in downtown Gloucester to make oilskins for fishermen, and as more relatives came from Europe to New York, some of them moved to Gloucester also. Morris's father came, too, and the family called him by the Yiddish name for grandpa, "*Zeyda*." But since few people in Gloucester could say "Zeyda," he became known as Shady—Shady Bell.

In 1909, Morris and a partner started Cape Ann Manufacturing, which made oilskins, other workers' clothing, and some fishing gear. His sons Richard and Harold took over the company during the Depression. During World War II, they made parachutes and Eisenhower jackets and later, joined by the youngest brother, Bradley, the company became a leading manufacturer of sportswear under the brand name "Mighty-Mac," named for their best-selling Mackinaw coat. Even when the company was no longer making clothes for the fishing industry, its logo remained the Gloucester fisherman.

THE CONSTANT APPEARANCE of newcomers, like the constant disappearance of family and friends, shaped Gloucester's character. The people of Gloucester were not always kind to newcomers, but their arrival was always expected. An old Gloucester joke is: What do you call someone who moved to Gloucester when he was a one-year-old and lived there to the age of ninety-nine? The answer: A newcomer.

Greeted with distrust and sometimes outright bigotry, the newcomers built their own churches, worked their own boats, huddled in their own neighborhoods, made their own foods, and changed the life of the town. One of the groups that permanently reshaped Gloucester culture was the people known along the New England coast as the "Portagee."

The fundamental thing that New Englanders do not understand about the Portuguese who settled there is that they did not come from Portugal. They came from the Azores, an island chain in a part of the mid-Atlantic known as the Saragosa Sea. The Azores are about a thousand miles closer to Gloucester than Portugal is, which makes them the closest piece of European land to Gloucester. They are culturally and politically part of Europe, but not on the European continental shelf. In fact, the sea around the islands is so deep that some of the mountains of the Azores, such as the snow-capped volcano that dominates the island of Pico, are the world's tallest mountains, if measured from the ocean floor where they begin.

When both the Flemish and the Portuguese visited these

islands in the early fifteenth century, they did not find a single human, or mammal, or even reptile inhabiting them. The Flemish and the Portuguese settled there, joined after 1492 by defeated Moors from Spain. This mixture, along with other European nationalities blended in later, have made Azoreans look different from the Portuguese, often considerably darker, and they speak different dialects of Portuguese, which vary from island to island.

The people of the Azores were largely agricultural. Many were wine producers, but a fair number were fishermen. They were also whalers and continued hunting whales until the 1980s. In the nineteenth century, whaling became a tremendously profitable enterprise in New England as well, and Gloucester got in on the craze, forming its first two whaling companies in 1832. Whaling was important to Gloucester because of spermaceti, a white, odorless, nonoily, waxy substance derived from whale oil. It is completely insoluble in water and was used to waterproof fishermen's clothes, which is why they came to be known as oilskins. In keeping with the self-sufficient tradition of the island of Gloucester, oilskins for the fishermen were made in town, as was the spermaceti, in a plant on Porter Street. It was owned by William Pearce, a merchant who sustained himself through periodic declines in the fishery with importing spices and sugar from Suriname and other nonfishing operations. In 1833 he financed a whaler out of Gloucester, but it did not have a particularly successful voyage. In fact, in New England even in the years when whaling profits were at their peak, ground fishing, especially cod fishing, brought in more money. And so Gloucester stuck to its traditional fishery, although by the mid-nineteenth century there were already profitable businesses taking tourists whale watching on the nearest banks, Middle and Stellwagen, between Cape Ann and Cape Cod, where slick, black giants—humpbacks and finback whales—squawked, grunted, puffed, hissed, leapt, and cavorted in plain view.

Whalers from the Azores would run across whaling ships from Nantucket, New Bedford, and Mystic, and they would sign on for better earnings. Once in New England, they would jump ship in towns such as Provincetown, New Bedford, and Gloucester to find better paying work. There was always room for new fishermen in Gloucester. From 1850 on, a steady flow of Azoreans came to Gloucester. Soon not just whalers jumping ship but whole fishing families, especially from the volcanic island of Pico, center of the Azorean whaling industry, and its neighbor, tiny Faial, having heard of Gloucester and its ground fishing, came.

In 1868 the Philadelphia-based *Lippincott's Magazine* ran a feature article on Gloucester, which called it "the most extensive fishing port in the world," and went on to say:

> The fishermen are a jolly set of fellows. Among them may be found all nationalities. The Portuguese form an important class, and there is quite a settlement of them in Gloucester. They come from the Western Islands, and are, for the most part, frugal, industrious citizens, fond of garlic, intense in their religious belief, which is Roman Catholic, and very superstitious concerning Friday, which they consider an unlucky day; and they will

never sail on that day if they can possibly avoid it. They are passionately fond of pictures representing Catholic Saints, and the walls of their dwelling houses are profusely decorated with such, very elaborately framed. They look upon them with feelings amounting almost to adoration, and indulge in the, to them, pleasing belief that these pictures possess power to bless and make them happy. They are very saving of their money, and, as soon as they get enough ahead, generally purchase a piece of land, build themselves a house, get married and open a boarding establishment.

They settled on a hill just east of downtown, a hill that reminded them of the Azores in the way it sloped down to the waterfront. The neat, closely placed, two- and three-story houses reflect the light from the harbor and glow in the late afternoon, which made the hill a favorite painting sight for the artist Edward Hopper.

It was and still is called Portuguese Hill, and in the late nineteenth century and for most of the twentieth century it was a Portuguese-speaking neighborhood in which English was not required to shop in local stores. The neighborhood residents made their own wine, planted kale in their gardens, and hung salt cod on their clotheslines to dry. A traditional meal always had one salt cod dish, usually the appetizer.

Alice Prady, who was born in Gloucester in 1898 of Azorean parents, lived on Portuguese Hill and made this dish as an appetizer or main course. The recipe was saved by her daughter, Mary.

BACALHAO SEBOLADÁ

> *1½ pounds salt cod*
> *1 cup ketchup*
> *2 tablespoons butter*
> *½ cup olive oil*
> *1 tablespoon hot red pepper sauce*
>   *(see recipe for linguica, p. 72)*
> *½ cup white wine*
> *3 large potatoes, sliced*
> *3 large onions sliced thin*
> *3 tomatoes sliced thin*
> *3 crushed garlic cloves*
> *1 large green sweet pepper cut into strips*
> *12 black olives, pitted and halved*

Soak the cod overnight in cold water, then drain. Boil the cod in fresh water for 15 to 20 minutes; rinse it in fresh cold water, and drain.

Mix in a small bowl ketchup, two tablespoons melted butter, olive oil, hot pepper sauce, and wine. Flake the cod and alternate layers of cod with layers of vegetables in a roasting pan. Drizzle catsup mixture over the top and bake for 1 hour in an oven preheated to 350 degrees.

For a town that today is associated with Roman Catholics and Catholic rituals it is striking that, in 1889, when the first Portuguese church was built on Portuguese Hill, it was only the second Roman Catholic church in Gloucester's history, the first having been built only eighteen years earlier. In 1914 the Portuguese church burned down. When it was rebuilt in 1915, The Church of Our Lady of Good Voyage was adorned with a model of a different vessel at each of the stations of the cross. On the roof, two great blue domes modeled after those on a famous Azorean church shielded a ten-foot statue of Our Lady of Good Voyage, an Azorean patron saint of fishermen, gently embracing in her arms a Gloucester fishing schooner. By the 1980s the statue had rotted and was replaced with a fiberglass copy.

The Portuguese church became the center of the Portuguese community and played a key role in the introduction of many Portuguese festivals. The Gloucester Azoreans added new ones unknown even in the Azores, such as the annual blessing of the fleet, which has become a tradition in various New England ports. In the Azores, the local priest blesses each new vessel as it is launched, but there is no annual fleet blessing.

In the 1890 census, the population of Gloucester was 24,651, of which 12 percent were Portuguese. But the Portuguese were soon to represent the largest group of fishing boat captains and crews, and would dominate the Gloucester fleet until the 1950s. The Portuguese fishermen encouraged their children to get an education and not to fish, so they began to vanish from the fleet.

Portuguese people are still coming to Gloucester from the Azores. But the Azores have changed. As a full member of the European Union, the archipelago today has little poverty and, for those in trouble, better social benefits than are available in the United States. Most residents of Portuguese Hill today do not speak Portuguese, and only one in five has a Portuguese background. Of those, many have intermarried, though they still teach Portuguese to their children and still attend Our Lady of Good Voyage.

Though there is no longer a single Portuguese-owned fishing boat in Gloucester, Portuguese culture remains part of the Gloucester way of life. They rarely cure salt cod on their clotheslines anymore, but they buy it from Canada for holiday meals. They still grow kale in their gardens and they start marinating pork the day before Easter to make *vinho de alios*. Their sugared bread, known here as Portuguese sweet bread, and their sausage, linguica, have become standard food in Gloucester, are served in local restaurants, are sold in shops, and are even in supermarkets.

Alice Prady's daughter, Mary, who had saved her mother's recipes, married Jack Gamradt, who despite a German father grew up on Portuguese Hill speaking Portuguese, as do many mixed-marriage Portuguese. His mother, Elvira Rose, was the daughter of Anton and Josephine Rose, who were linguica makers in Lages, Pico Island, in the Azores and who moved the business to Gloucester. This is their recipe:

LINGUICA

> *1½ cups cider vinegar*
> *1½ cups red or white wine*
> *3 cups water*

12 pounds boneless pork butt, ground so small pieces are
    about ½ inch thick
sausage casing, rinsed
4 tablespoons salt
3 ounces peeled and ground garlic
¼ cup paprika
¼ cup lemon juice

Mix marinade together in a stainless steel bowl. It must be stainless steel. Add the ground pork and mix well. Cover and refrigerate for 24 hours. Then stuff it into a hog casing. In a smokehouse (use hickory), hang the linguica on wooden poles until it is dry to the touch and the color looks nice. It's done.

JOHN SMITH HAD BEEN wrong when he speculated that Cape Ann might be a good place to make salt. That is why Sicilians had been coming to Gloucester almost as long as the Azorean Portuguese. Before refrigeration, fisheries such as Gloucester needed huge quantities of sea salt to cure their catch for market. Sea salt was made, and still is, by evaporating seawater in the sun, which requires a sunny climate. Each rainfall is a setback. After a year or more in the sun, a pool of seawater produces large crystals, because the more slowly crystals are formed, the larger they are. It is the size of the crystals, not the fact that it comes from the sea, that makes sea salt essential for curing fish.

Gloucester is too far north for producing sea salt, but the Mediterranean is ideal for salt making, and Trapani, on the western coast of Sicily, is one of the oldest and most important saltworks in the world. The Phoenicians made salt in this spot for curing tuna to trade and later the Romans did the same. Salt was also brought to Gloucester from Salt Key, a tiny Caribbean island in the Turks and Caicos that produced nothing but salt and even paved their few roads with it. Occasionally salt came from Cadiz, in southern Spain. But the Gloucester salt cod producers found that the type of salt used made a huge difference. Some salts tinted the fish slightly orange, and such a fish was unsaleable.

The Trapani salt gave the most consistently good results. Not all the ships that brought it were Sicilian, though most were. Others were Portuguese or Spanish, or even English or Norwegian. They were barks—huge, wooden-hulled, two-masted, square-rigged vessels—with broad, rectangular sails that were easy to spot as they approached Eastern Point. They sailed into Gloucester Harbor and dropped anchor in the deep part, because the barks sat very low in the water when their holds were full of salt. Smaller vessels called "lighters" would go out and unload the bark until it sat high enough in the water to sail up to a wharf to finish the unloading.

The salt barks spread the news in the fishing communities of western Sicily of this booming fishing town in America called Gloucester. Sicily was one of the poorest places in Europe, and since the late nineteenth century, Sicilians had been migrating in search of a place to earn a living. They moved into the Fort and made it a crowded little jut of land in the

harbor, where they lived together in wooden houses with peel-
ing paint and walked to work, the fishermen to the nearby
docks and their wives to the neighborhood fish processing
plants.

The salt barks from Sicily came until the 1930s, when
frozen food killed the salt cod business. Joe Santapaola, in his
nineties, still remembered working on his father's lighter.
"Down in the hold we shoveled salt into half hogs-head tubs. A
hogs head was 1,100 pounds. We hoisted them from the hold,
one going up all the time. It took two and a half hours to unload one hundred tons."

His father's lighter, because it had a derrick for hauling salt,
also replaced masts on other boats. Santapaola rescued a spar
that his father had replaced on one of the Sicilian barks. His father used such scrap for firewood. "It was eighteen inches in
diameter and not a knot in it," said Santapaola. In 1940 he
spent a year carving the yardarm and fashioning it into a four-
foot-long model of a Grand Banks schooner with operating
rigging. One thousand solder joints for handmade brass fittings and blocks served the hull with a lead keel. He intended it
to be sailable, but he put it in the water only once, in his bath-
tub, to see if it would float at the waterline. "It was right there,"
Santapaola said with a slight show of pride. He displayed the
boat in his otherwise modest East Gloucester living room.
His basement was also full of models. He made six 27-inch
schooners for his six children and then twenty for twenty
grandchildren. He also built other models, such as a tugboat
with a working four-cylinder steam engine. And he even built a
few to sell. But few outside of his family knew of this sideline in

his long career as a commercial lobster fisherman. Gloucester homes are full of such surprises.

In Santapaola's youth, salt cod was the leading business of Gloucester. Santapaola remembered how workers at the fish flakes scrambled to put the fish in covered boxes anytime there was a sign of rain. Huge fish companies such as Gorton's bought up the catch.

In the 1840s a wholesale fish merchant called Fluid Smith had started skinning and boning cured cod and selling it as "salt fillets." Slade Gorton, an out-of-work superintendent from one of the cotton mills that prospered in eastern Massachusetts at the time, visited from Rockport. After studying Fluid Smith's operation, in 1849 he bought salt cod with his life savings, filleted it, and packed it in wooden boxes. The little boxes quickly became Gloucester's most popular product and Gorton established plants and fish-drying flakes. The company further expanded under his sons, and in 1906 merged with Gloucester's largest fishing fleet, owned by John Pew, to form Gorton-Pew, the largest seafood company on the Eastern Seaboard of the United States.

This early twentieth-century recipe for boxed salt cod, a kind of cooking that today is rarely found in Gloucester, was published shortly after the merger of Gorton-Pew Fisheries:

> Tear the codfish and set it in a pan of water on the back of the stove: let it simmer, changing the water several times, until the fish is properly freshened. Let it boil five minutes. Never use lard but cut clear, fat salt pork into half-inch cubes and fry over a slow fire. When you

Established 1849.

JOHN PEW & SON,

Producers of and Wholesale Dealers in

FISH,

And Importers of

SALT,

83 Spring Street, Gloucester, Mass.

JOHN PEW.   CHAS. H. PEW.   JOHN J. PEW.

An 1876 advertisement showing a Gloucester fishing schooner

take up the fish, have some boiled potatoes and beets ready, and make a hash of three-sixths fish, two-sixths chopped potatoes, and one-sixth chopped beets: turn the pork fat into the hash. Mix thoroughly and fry brown. Boiled onions will make an excellent side dish for this dinner.

A giant in the salt cod trade that spread its salted fish for drying throughout the town's waterfront and provided the wives of immigrant fishermen with jobs curing and taking care of these fish, Gorton-Pew had an insatiable appetite for Sicilian sea salt to cure all this fish. But in 1929, Clarence Birdseye sold his quick-freeze process to General Foods and in 1930 they started selling frozen food. "The salt business died when frozen fish came along," recalled Santapaola. "It quickly did away with salt in a few years." In 1950, fishing captain Tom Benham, interviewed by a Gloucester newspaper about the disappearance of the salt barks, said, "The young people today just don't want to eat salt fish. In fact, I don't either."

After World War II there was a new wave of Sicilian immigration. During the war life had been hard for Italian-born fishermen in Gloucester because they were not permitted to go to sea or work on the waterfront. But after the war there was opportunity. Gloucester and other American ports had a shortage of fishermen because the men had gone to war and came home to the GI Bill offering them education and new careers. Sicilians, whose always-weak economy had been devastated by World War II, went to fishing towns not only in New England but also in the Gulf, the Great Lakes, and on the West Coast.

While the first Sicilians in Gloucester had been from Trapani, this new wave, still from the western half of the island, came from the north coast, from Terrasini and five other fishing towns near it. In 1957, the poet Charles Olson moved into the Fort, a rare non-Sicilian in the neighborhood. He called the Fort "Algonquin-Sicilian" because it seemed to him to be a tribal society. Not only was everyone Sicilian, they were mostly from Terrasini.

A characteristic of that part of the Sicilian coast is that there is very deep water almost to the waterfront, so fishermen did not have to go far for a catch. They used a purse seine, a technique in which a school of mid-water fish—in this case anchovies and sardines—would be surrounded by a net. The ends of the net were then hoisted up and drawn together like a purse, and the net with an entire school of fish was towed by long rowboats, hauled to shallow water where the fishermen scooped the fish out. This led to a tradition of racing these long rowboats on holidays, and today in Gloucester the seine boat races are still a major feature of the St. Peter's festival.

For tuna—bluefin, yellowtail, and bonito—the fishermen of Terrasini would drag a net between two boats, a technique known as pier fishing, almost in front of the town. Occasionally pandemonium would break out as a swordfish caught in the net went berserk, a spear-headed maniac, water foaming, net straining, men shouting, while the townspeople cheered on from the shore.

A seine boat

Terrasini is just west of the Sicilian capital, Palermo, today near the airport. There is evidence of inhabitants dating back to Paleolithic times in the striated caves carved by the sea into sheer limestone cliffs—the grottos. But fishermen only began working in this essentially agricultural village about the time they did in Gloucester, in the seventeenth century.

Considering the hundreds of people it contributed to Gloucester it is striking that Terrasini has only about ten thousand residents. Today many of the last names of Terrasini are the last names of Gloucester. There are sixty-nine separate listings in the Gloucester telephone book for the name Ciaramitaro. Some of the Gloucester Ciaramitaros don't even know each other. They know only that their ancestors all came from Terrasini and so they might be related. The same is true of the fifty-seven Fronteiro families listed in Gloucester.

In Gloucester, Sicilians met with considerable prejudice— far more than most new groups. If this was because of their closed "tribal" way of living, as some non-Sicilians have suggested, or if they were closed because they encountered prejudice, as some Sicilians have suggested, is uncertain. They spoke a language that was not understood even by Italians—not just Sicilian but western Sicilian, a local dialect from the Terrasini area, a dialect of a dialect. Sefatia Romeo said, "It's nice to know a private language."

Lena Novello, born in 1915 into a Sicilian fishing family, recalled their first years in Gloucester: "It was hard for them when they came; they couldn't speak the language. The people around them wouldn't communicate with them. . . . They used to call them Guineas." She remembered the other fishermen being unwilling to share vital information with the Sicilian fishermen. "The men would go trawling to start and they didn't know where there were shipwrecks. They would lose gear, no access to information, no one wouldn't tell them anything. Because they didn't want these Italian people around. They wanted them to go away."

Harold Bell recalled his father's factory in the Fort. "My sister was not allowed to come down because there was this feeling about Italian men. The Babson Library never had a woman working alone because of fear of Italian men."

But this treatment only drove the Sicilians closer to each other. They lived together, sometimes more than one family in an apartment, bound by the same hard life that had always bound Gloucester people. Tragedy was a way of life. The large Sicilian families tried to get several boats so that if one went down the entire family wasn't lost. "There wasn't a weather report," said Lena Novello. "You would get up at night to see how the weather was. My father tied a white handkerchief on a clothesline, during the night he'd get up and he'd look. Oh, I guess it's blowing bad weather. I won't call the boys—his sons—to go fishing. I remember we would go up and play in the Scout Mountains, and we would see a boat coming in half-mast and we would all be petrified and we'd all run home wondering who it was that got lost at sea—I remember, I wonder if it's my father, I wonder if it's my brother and we would all run down. And after you find out who it is and everyone says, 'Oooh, who's going to go tell this to that family?'"

In 1969, Lena Novello became the first leader of an organization for fishermen's wives that was one of the most powerful and assertive political lobbies in Gloucester. The group also produced a spiral-bound cookbook of largely Sicilian recipes that to everyone's surprise, sold 100,000 copies. In 1977, after Lena Novello's death, Angela Orlando Sanfilippo, a five-foot-tall Sicilian-born woman with a hoarse and angry voice, so passionate that she frequently seems on the verge of tears, took over the organization with relentless energy. The successes of the Gloucester Fishermen's Wives Association, largely though not exclusively Sicilian, most notably in working with environmentalists to stop a 1979 plan to drill Georges Bank for oil, helped to convert the Sicilians from outcast immigrants to civic leaders.

Angela Orlando arrived in Gloucester as a teenager from Porticello, a town near Terrasini. She learned English and translated for her fishing family. That was the way it was done; one of the children would learn English and become the family translator. Angela was an honors student at Gloucester High School, but her parents would not allow their daughters to go to college. They had work to do for the family.

In August 1963, when she was still a child in Porticello, the fishermen were sitting outside as they always did in the evening. Her grandfather came running out and told everyone to go indoors because the end of the world was coming. They all huddled in their houses as a violent storm tore through Porticello, and Angela thought she could hear women shrieking in the distance. She remembered a teenager, John Sanfilippo, coming back from sea and reporting that his two brothers had been off jigging in a small dory and had been caught in the storm and drowned. But later the brothers came back and lived to go to Gloucester. Angela always remembered that strange evening, and years later when she moved to Gloucester and saw the same John Sanfilippo, she was certain they were fated to be together and they married. Sicilians often believe that fate is the force that steers lives.

EVEN LIVING ON THE threadbare side of a ramshackle industrial port, the Sicilians felt that they had arrived at a wondrous place. Tom Biacaleone, whose father had been a captain and civic leader in Terrasini, arrived in Gloucester in 1956, days before his thirteenth birthday, with his mother and five younger sisters. The father had come two years earlier and had established the family fishing business. Tom, who had never seen a television or watched a movie, still remembers, "I thought Gloucester was like the nativity scenes we made in Sicily, the way each separate house had a little green around it."

The Biacaleones borrowed money and bought boats with the help of a family member who had been in Gloucester since the 1920s. They operated three fishing boats almost entirely crewed by family members. Tom recalled there being as many as five thousand Sicilians living close together in the little Fort area. Many lived without hot water, some without any running water at all. Children went to the docks and begged a few fish to sell in town for a nickel each. But Biacaleone remembered the crowded dilapidated neighborhood fondly as a home. "You didn't feel like you were lost," he said.

Francesco Ciaramitaro was born in 1908 in Terrasini, but

like many poor Sicilians, he migrated to northern Italy in search of work. He did not know his parents because they went to America and left him in the Piedmont with an aunt known as Nonna Maria. For Francesco Ciaramitaro, finding his mother became an obsession. At the age of sixteen he started going to every port in Italy and looking for a way to stow away on a U.S.-bound vessel. This was mostly during the rule of Mussolini, when Italian justice was brief and brutal, and Francesco was regularly arrested and beaten. For thirty years he tried to get on a boat to America. Even after he married and had children, he kept trying.

Finally, in 1955, he got a job as a cook on a freighter and he left his wife and eight children. He thought he was going to America. But the freighter went to *South* America, then South Africa, and India, before finally heading to Los Angeles and then through the Panama Canal to New York. In New York, he jumped ship—literally, he jumped over the side into New York Harbor and swam to shore. Soaking wet in lower Manhattan, not knowing a word of English, he wandered until a taxi driver stopped and said in Spanish, "Hey what are you doing?" and bought him a cup of coffee.

He found that his mother had died eight years earlier. No one had told him. He had two brothers and a sister in America, in Gloucester, and they had not contacted him. Francesco joined them in Gloucester and started fishing because that was what you did in Gloucester, and he could earn enough money to send some back to his wife and eight children. This continued for two years, and then someone turned him in to the Immigration and Naturalization Service and he was deported.

Eleven years later, Francesco returned to Gloucester. This time he took his wife with him and, in the Sicilian way, he also brought his youngest child, twelve-year-old Paolo, as a translator. This time they entered legally through Ellis Island. "The Statue of Liberty was so big!" Paolo still related with excitement years later as an adult.

In Gloucester, Francesco washed dishes at the Captains Courageous, a restaurant named after the novel of Georges Bank fishing that Kipling had written while living in Gloucester. Francesco earned $80 a week and Paolo washed dishes part-time for $40 a week of five-hour shifts while attending Gloucester High School, all the time remembering a Sicilian girl, Maria Anna, he had met in the Piedmont two years before leaving for America, when he was ten and she was five.

Paolo wrote letters to Maria Anna, and in 1974 he became a U.S. citizen. The following year he returned to Italy for an official Sicilian-style betrothal. In 1978 he returned with his wife, who he supported for the next twenty years as a fisherman on bottom draggers.

"I never liked it," Paolo said about fishing.

In 1989, he and Maria Anna returned to Italy, where they studied pastry making for three years. Then they returned to open Caffè Sicilia, a Sicilian pastry and espresso bar on a corner of the curved, shady, brick storefront, Norman Rockwell–perfect, New England Main Street where it fits in perfectly with numerous other Sicilian establishments.

Cannolis, those dark, tubelike Sicilian cookies filled with sweetened ricotta cheese have also become very Gloucester— as Gloucester as linguica. Paolo's cannolis are exactly like the ones in Sicily, except bigger. Paolo is bigger, too. Gloucester Sicilians like things big, and their cooking is a last bastion against *cuisine minceur*. Take this cooking tip from Rosaria Giambanco's hair salon. Rosaria was giving a recipe to another Sicilian woman:

"Then you add the cheese."

"Mozzarella or American?"

"Both!" came Rosa's bold reply.

"Wow!"

"Yeah, then I like to add grated cheese on the top."

"Wow!"

"Yeah!"

This is Paolo Ciaramitaro's favorite pasta recipe for four people:

> *Slice a little fresh hot pepper*
> *Chop Italian parsley—"Italians don't use that curly*
> *garbage."*
> *Chopped mint*
> *4 vine-ripe round tomatoes, chopped*
> *8 tablespoons extra-virgin olive oil*
> *Garlic*
> *A piece of calamari, a piece of cooked lobster,*
> *and 10 little shrimp*
> *Pasta*
> *Mozzarella, Romano, and Parmigiano cheese*

Sauté in the olive oil; it's got to smell like the garlic. There has to be a lot. Penne, rigatoni, or spaghetti cooked *al dente.* Drain, and save a little of the starchy water. Mix everything and add water a little at time. Chop mozzarella in cubes. Put it on top with Romano and Parmigiano.

Gloucester Sicilians take pride in clinging to Sicilian ways more than most of the Sicilian communities in America. This is probably because it is a community of fishermen, and in all societies fishermen are notorious for preserving traditions. They look forward to a long list of holiday events, most of which involve food. St. Peter and St. Joseph, the two patron saints of blue-collar labor, are revered by Sicilians. The evening before March 18, St. Joseph's Day, the Sicilians visit the various Sicilian altars around Gloucester, and they eat an orange, a lemon, and a piece of bread. The orange is for a sweet future, the lemon to remember the suffering of hard times in the past, and the bread so that you may never go hungry, and, according to Sefatia Romeo, to remember that there is still hunger in the world.

For All Saints' Day in November, children find new shoes behind their door filled with cookies. Every New Year's Day there is pasta with lentils. "It has to be handmade pasta," Paolo Ciaramitaro cautioned. Fennel, broccoli, cauliflower, and chickpeas are added. For Christmas there are always fig cookies. This is Ciaramitaro's Christmas fig cookie recipe for his Caffè Sicilia:

> *1 pound cake flour*
> *1 pinch baking soda*
> *1 pinch baking powder*

1 pound high-gluten flour
1 pound pastry flour
1 pinch vanilla extract ("a pinch is a teaspoon")
1 pinch lemon extract
1 pinch orange extract
8 eggs

Mix it all in the mixer at medium speed. Add the flour last. Roll out the dough. Grind green Greek figs. Add sugar, honey, and a pinch of lemon zest and a pinch of orange zest. Mix it. Add the figs to the dough. Roll it into a log and bake at 325° in a gas oven or 355° in electric until the crust is brown. Cool and slice it.

For Christmas, too, there are seven dishes symbolizing the seven sacraments and another thirteen dishes for the apostles. All of the seven dishes of the sacrament are fish, including lobster, squid, octopus, gray sole, and salt cod. There is no meat until after midnight. While the Portuguese, who no longer fish, buy their Christmas salt cod already cured from Canada, the Sicilians take Halloween as a cue to bring home a codfish and start curing it for Christmas *zuppa di baccalà*. Here is Paolo Ciaramitaro's recipe:

Coat the baccalà in flour and fry it in olive oil. Caramelize onions and vinegar with it. Bake it with two tomatoes, capers, carrots, potatoes, celery, and olives on top of the stove.

During the 2006 soccer World Cup, when the United States played a very tense game against Italy, it was clear that, among the men sitting around the television at the Caffè Sicilia, sipping espresso and shouting in Sicilian, there was nobody rooting for the United States. Their culture seemed to be safe on the island of Gloucester. There, they could be as Sicilian as they wanted and still consider themselves flag-waving, 100 percent Americans because to the rest of Gloucester, Sicilian, like Portuguese, had become "very Gloucester."

chapter Six

AMONG THE ROCKS

...

Come among the rocks then, solemnly and speak
not even to the wings that pass in flurry
to encompass earth and sea, and other windwing worry;
I sit a spell, clutching at plain thoughts, wrenching at
no secrecies, hearing the magnificat
of afternoons and mornings united in their theme . . .

—MARSDEN HARTLEY,
"SOLILOQUY IN DOGTOWN," 1931

GLOUCESTER'S POPULATION GREW IN FITS. WHEN THE FISH-
ing was good it increased rapidly. During times of "fishing
crises," the size of the population would barely change.

The early settlers lived along the coast. They regarded inland tracts of land as common land and locals were all given
timber rights to this interior forest. As the land was cleared
they were given common grazing rights. The area was too
rocky to be desirable farmland. According to popular Gloucester folklore, this central swampy area was the last place on
earth to be made, and so all of the world's unused rocks were
dumped there. Some of these rocks are gargantuan domes or
blocks of granite ten feet and more in height.

If not the last place on earth to be made, at least it was the
last place in Gloucester to be settled. In the beginning of the
eighteenth century, with 650 people living in Gloucester, inte-

rior lots of cleared land—common grazing land—were offered to any male who had reached the age of twenty-one. By the mid-eighteenth century, twenty-five families, including some of Gloucester's leading citizens, had settled in the center. Fishing was in one of its chronic periods of decline, and Gloucester people were looking inland. Fishing became even worse as war approached; navies always impede fishing in times of war. But after the American Revolution, fishing boomed again and the population grew dramatically along the coastline, especially near Gloucester Harbor. One by one, the homes in the interior were abandoned. Only the poorest people of Gloucester remained there—the area became known as Dogtown. Some in Gloucester say that the name comes from the many dogs that were abandoned there; others say that it refers to the many dogs that lived there during the Revolution to protect the women while the men were off to war. Helen Mansfield, an expert on Cape Ann dialects, believed that the name derives from some other now-lost phrase that gradually, through mispronunciations, turned into "Dogtown." But Cape Ann historian Joseph Garland pointed out that there have been other Dogtowns in the area and that the name usually denoted a slum.

Dogtown, a Gloucester story in itself, is the site of other Gloucester stories—stories within stories. In 1830 the last resident of Dogtown was taken to a poorhouse, and in 1845 the last house was torn down. The slum in the center of Gloucester became an abandoned area, its cleared fields quickly taken over by thick, secondary growth with only its huge rocks remaining bare. "We could see no houses, only hills strewn with boulders, as though they had rained down," wrote Henry

David Thoreau of his 1858 visit to Dogtown. Strange things went on in the part-swamp, part-woods, pocked with occasional holes beneath brush that had once been cellars of now-vanished wooden houses. The area became a place for eccentrics, something of which Gloucester has never had a shortage. In 1892, a six-foot-seven-inch-tall fisherman named James Merry decided at the age of sixty to become a matador and he raised a calf to be a fighting bull. He worked out with him in the secrecy of Dogtown. A sign on a rock marks the spot where Merry was killed by the bull.

In 1920, Marsden Hartley, a well-turned-out forty-three-year-old dandy but also one of the few artists to have garnered equal reputations as a major painter and an important writer, came to Gloucester and discovered Dogtown. Later when Hartley, who was born in Lewiston, Maine, became consumed with painting and writing about New England, he remembered Dogtown:

> I had seen some rocks twelve years before that when I was in Gloucester and Gloucester is one of those you want to go back to—I had remembered the rocks and the name Dogtown—that's a great name.

The drawings, paintings, and poetry he produced there in 1931 and 1934 are considered among this master of modernism's most important works. Dogtown remained an overgrown wilderness. A 1922 attempt to replant it as a forest had failed. While drawing in Dogtown, Hartley came across a "Scandinavian" stonemason carving "some economic mottos" on Dogtown boulders. The man said he worked for Roger Babson, a wealthy local who had purchased a large tract of Dogtown.

Roger Babson, born in 1875, had herded cows through the paths of Dogtown as a boy. A pioneer financial analyst, he made a fortune, founded three colleges, and supported various causes. He once ran for president on the Prohibitionist Party ticket. He was the kind of man who would lecture people whether they listened or not. And he loved Dogtown. He published a guidebook with a map and posted numbers at points of interest indicated on his map.

Although Babson expressed outrage at those who marked up Dogtown with graffiti, during the Depression he hired thirty-six out-of-work Finnish quarry workers to carve his pet

Dogtown

moral precepts on the massive boulders of Dogtown. They remain and as the hearty hike through the rugged landscape, they still pass rocks with mottos on them, including "Save," "Be Clean," and "Get a Job." Marsden Hartley was not the only one to protest this defacement. Babson's angry family said that he had disgraced them.

Today Gloucester has a protected wilderness in its inland center. Dogtown is a five-square-mile patch of boulder-studded woodlands and swamps with two reservoirs for the city's water supply, trails, abandoned roads, and the remnants of cow paths. It is the home, naturalists say, to five thousand plant species but to no one else. Development is forbidden. Depending on one's point of view, it is either a protected wilderness or an overgrown ghost town, returned to nature, in the center of Gloucester.

IN 1840 THE SIZE and population of Gloucester suddenly declined as the northeastern tip of the Cape was carved off into a separate town called Rockport. This area, according to the breakup agreement, measuring two-sevenths of the area of Gloucester, was one of the last to be settled but had the fastest-growing population. In the half-century between 1792 and the split, the population of Gloucester had doubled, but the population of the Rockport area had quadrupled. It did not have Gloucester's "Beau Port"—its extensive sheltered harbor in the lee of the wind—but over on the windward side of the Cape, instead of the single-pronged economy of Gloucester, Rockport had three branches. As in Gloucester, there was fishing, especially after an artificial harbor was built in 1811. But because it was not deep and sheltered, the fishing never reached Gloucester's scale, so there was also a serious attempt to make agriculture profitable. And there were the quarries. It was these quarries, which mined and shipped Cape Ann granite, that gave Rockport its name and a different character from Gloucester.

Thanks to the glaciers that carried the rock and to the fault line that trapped it, Cape Ann is one of the few places on earth where such hard and dense granite is found at the surface. The speckled rock comes in a variety of colors. Light gray with charcoal speckles is the best known, but there is also rust, yellow, black with white speckles, green, orange, and brown.

Historians argue about who brought stone-cutting techniques to Cape Ann in the eighteenth century, but several quarries were established at the time, most notably one on Halibut Point. Like fishing, quarry work was hard physical labor requiring great skill. A stone cutter marked a stripe on a block of granite with a chalk line and then cut a groove along this mark with a hammer and chisel. He then pounded metal spreaders into holes along the line and carefully drove rivets into the spreaders until the rock split along the line. Since the quarries had the great advantage of being by the sea, the heavy stones could be loaded onto sloops and transported by water. But in 1879 a rail line was built to transport large pieces.

The biggest market for this hard granite was as paving stones for city streets. Boston, New York, and Havana were all paved with Cape Ann granite. The Brooklyn Bridge was built out of it, also.

Like fishing, stone cutting was dangerous work. White lung disease from breathing granite powder was common, and few quarry workers were able to continue beyond the age of forty. It was a different life from fishing and it drew different immigrants, especially from Finland. These were not individualists working in the wilds of nature for a share of the take but badly paid wage laborers.

In October 1929, the quarry workers' union called a strike. Soon after that the Stock Market crashed. Most of the quarries never reopened. Halibut Point was left exactly the way it had been when the workers put down their tools and walked off the job. Some of the pits have filled with rainwater and become favorite deep-water swimming holes.

IN 1870, GLOUCESTER'S fishery was prospering, and this led to a growth in population. In the five years since the Civil War, the population had increased by 3,602 to 15,397, which made it the largest municipality in the state of Massachusetts to be called a town. In 1873 a new charter was approved, officially designating Gloucester as the City of Gloucester. The population has not even doubled to modern times, and today, at 29,000, seems small for a city. But it is an odd point of either pride or amusement, depending on how you interpret it, that if you refer to Gloucester as a town, locals will correct you.

The city is somewhat more populous than it appears on paper because there are many part-time residents. As early as the 1820s affluent Bostonians began looking to the coastline directly north of their city, the area that was to become known as the North Shore, for sea air and summer relaxation. Their first resort area was the town of Nahant just north of Boston. In a few years this barren, seaside pastureland became the site of hotels and fashionable visitors, including the poet Henry Wadsworth Longfellow. During the nineteenth century, wealthy Bostonians gradually moved farther up the North Shore every summer, eventually making their way out to Cape Ann. In 1847 a railroad connected Boston to the North Shore, including the island of Gloucester. This railroad was the first of several links that were to threaten Gloucester's isolation. But while the rich Bostonians went to Beverly, then Manchester-by-the-Sea, then Magnolia, then Rockport, they hesitated to invade Gloucester-by-the-smell—as an old joke went. Robert Grant, a Boston Probate Court judge and popular author, wrote in 1894 "civilization properly ceases before you come to Gloucester."

An exception was Eastern Point, a narrow protrusion of land in East Gloucester at the opening of Gloucester Harbor. The area remained farmland until 1889, when a developer planned to cover it with houses. After construction was completed on only eleven houses, he ran out of money, mainly because of the cost of building a road through Eastern Point. But these few stately homes had a view of the rocks where egrets grazed and cormorants dried their wings and fed on the baitfish driven in by hungry striped bass.

These summer residences were marked by what was politely called "exclusiveness," which meant exclusion by race, ethnic origin, and income. The homeowners could watch fishing boats sail out and home again, a picturesque commercial activity with which most of them had no connection. Most of Gloucester could not qualify to frequent The Hawthorn Inn in East Gloucester or the Oceanside in Magnolia, or later the Eastern Point Yacht Club or the Bass Rocks Golf Club. This concentration of wealth on the waterfront was a new thing for Gloucester. The water was industrial and smelled of the fish that was dried along the shore. The more affluent people of Gloucester built on the hilltops with commanding views of the harbor but away from the smell. Now on the eastern side of the harbor, a safe distance from the fish flakes where salt cod was drying, hotels and homes for the wealthy spread out to Eastern Point, including one of the largest luxury hotels of the North Shore, East Gloucester's 300-room, five-story Colonial Arms.

Most of the summer people—the wealthy in their summer homes, resort hotels, and private clubs—had little connection with life in Gloucester. Harold Bell, who grew up in downtown Gloucester, said:

> The summer people were a whole other class with their chauffeurs and boats. They came with their trunks. They had the Bass Rocks Golf Club for their cronies. No Catholics. No Jews. Just cronies who came for the summer. The only time we had any contact with the summer people was when we collected money for the hospital or the YMCA.

The no-longer-standing Pavillion Hotel, Pavillion Beach, circa 1870
(COURTESY OF THE CAPE ANN HISTORICAL ASSOCIATION,
GLOUCESTER, MASSACHUSETTS)

In 1926, John Hays Hammond Jr., born in 1888, the son of
a famous and wealthy mining engineer, recalling the great es-
tates he had loved when living as a child in England, began to

build a castle complete with drawbridge in West Gloucester as a present to his bride. Among its unique features is a ten-thousand-pipe organ.

A child prodigy inventor who was nurtured by friendships with both Thomas Edison and Alexander Graham Bell, Hammond got his start as a teenager at the Lawrenceville School in New Jersey, when he foiled the dormitory's lights-out-at-eight rule by building a sensor into his door that would shut the light off if anyone opened the door. He made his reputation from his lordly domain in West Gloucester, where he developed radio remote control, sending unmanned vessels zipping around Gloucester Harbor. In 1914 he sent an unmanned yacht from Gloucester Harbor to Boston and back, a 120-mile voyage. With World War I erupting, he invented a system to protect his remote signals from enemy interference, a system by which a weapon could electronically seek out a target, and worked on the first radio-guided torpedo. Other inventions included an eye wash, a harmonic organ, an altitude-measuring system for airplanes, a magnetic bottle cap remover, a meat baster, a "panless" stove on whose aluminum surfaces food was cooked. His cure for baldness was one of his rare failures.

While Hammond was piloting unmanned boats from one side of the harbor in this scrappy, blue-collar fishing town, Clarence Birdseye, another genius inventor, the man who revolutionized the food industry and the Gloucester fishery with his fast freezing process, lived on the other side of the harbor in an ample brick home in Eastern Point, and built his fish-freezing plant in the Fort on the downtown waterfront. Although Birdseye's invention of frozen food ended the Gloucester salt cod business, it created the frozen-food business, which was equally profitable for Gloucester. Until his plant closed in 1965 it was one of the largest employers in Gloucester.

When he died in 1956, Birdseye had 250 patents to his name, mostly for food-related inventions, but also for an electric sunlamp, an improved lightbulb, a whale harpoon gun that placed identification markers in finbacks—chasing whales was among his many hobbies—and a process to make paper from the pulp of crushed sugarcane. But even this astonishingly prolific outpouring comes in behind the estimated 800 inventions and 437 patents of Hammond's.

AND THE PAINTERS kept coming to the famous rocky coasts, harbors, and seas of Fitz Henry Lane and the Luminists. The year Gloucester became a city, 1873, was a tragic year in which 174 men were lost at sea. But it was also the year that Winslow Homer, a thirty-seven-year-old illustrator known for his depictions of the Civil War for *Harper's Weekly*, a leading publication of the nineteenth century, came to Gloucester. The Boston-born Homer was at least well-known enough then for the *Gloucester Telegraph* to comment on his arrival and on his illustrations of Gloucester that were to run in *Harper's Weekly*. Gloucester, a town where the people are bound together by tragedy, has always welcomed anyone who might bring it fame, glory, or a little recognition from the outside world.

In Gloucester, Homer turned from his illustration work to watercolors. By then, great advances in the quality of watercolor paints had been made, and in 1875, two years after his first visit to Gloucester, the first portable travel kits of water-color paint became available. At the time, watercolors had no standing at all in American art. It was Winslow Homer in Gloucester who changed that.

Between his first and second visits to Gloucester Homer enhanced his reputation with oil paintings, many on Glouces-ter themes. He was fascinated with the subject of boyhood, and a series of paintings of Gloucester boys heeling to the wind in a small catboat did much to make him a major American artist. Homer's watercolors, which were sketched and then painted in, were not well received by critics, who found them imperfect and unfinished. In truth, the critics were simply not used to the idea of watercolors. In America in the late 1800s, a major artist, by definition, did oil paintings.

In 1880 Homer returned to Gloucester. He had abandoned magazine illustration five years earlier and had made up for the decline in income by painting hundreds of watercolors in up-state New York, Long Island, Maine, and in Virginia, where he depicted freed blacks. In the wake of the Civil War, America was rapidly changing, and Homer, well aware of these enormous shifts, liked Gloucester for its timelessness. But the city of fish-ermen harvesting the sea from schooners was already experi-encing many changes, he being one of them. The old Gloucester did not have luxury hotels or sumptuous summer homes or vis-its from major artists. But the new Gloucester still looked very much the same as it had before the Civil War, and there were ever fewer places in America that could make that claim.

For reasons that are not clear, by the time Homer returned

to Gloucester in the summer of 1880, he had become a recluse and he settled on tiny Ten Pound Island, off Rocky Neck, a bit of land and rocks in Gloucester Harbor that even today remains uninhabited. He lived in the lighthouse, rowing the short distance to shore when he needed supplies. The isolation seemed to work well for him, because he produced there some of the most stunning watercolors in American art. From Ten Pound Island and that part of Gloucester Harbor, especially in the summer, the light and the colors—the purples, reds, and ambers—are so dazzling they can leave people speechless, and it was this that Homer captured. To understand the beauty of this grubby fishing town and why watercolorists have been coming here ever since, one need only look at Winslow Homer's 1880 Gloucester watercolors.

Gloucester people were starting to like the idea of having famous artists in their midst. Another was William Morris Hunt, who introduced to America the Barbizon School, a new brand of French realism centered around the French village of Barbizon near Paris where such painters as Corot, Rousseau, and Millet gathered. They portrayed landscapes realistically without the drama of the Romantics, and their work was a precursor to Impressionism as well as Realism.

Born in Vermont and raised in Connecticut, Hunt made his reputation as a portrait painter in Boston. After having lost much of his life's work in a studio fire in Boston in 1872, and after his wife and children left him in 1875, he arrived in Cape Ann in 1877, a famous but sad and ailing figure. He settled in Magnolia, the little village just west of the Annisquam in the swamp where the flowers that gave it its name grew. The neighbors said that the new resident was "a great artist and his coming to Magnolia would be the making of the place."

A noted architect, though not as noted as Hunt's own brother, who designed the base of the Statue of Liberty, William Ralph Emerson, converted a barn into a studio for him, which Hunt called "the Hulk." Rather than live on a deserted island in the harbor, Hunt created isolation for himself by furnishing the Hulk with a gazebo on a tall pole that was accessible from the roof of the Hulk by way of a drawbridge. The gazebo was high enough to offer a view over the treetops to Magnolia Harbor, and Hunt could go there, pull up the drawbridge, and paint.

Hunt also had a covered wagon, which he claimed was built by a maker of gypsy caravans, drawn by two horses and equipped as a moving studio. He traveled around Gloucester, and when he found a suitable spot he would jump off the wagon. While his assistant set up an easel and paints and brushes, he would sketch out the scene in charcoal.

Hunt opened the way for other European-influenced painters, especially American Impressionists such as Childe Hassam, who painted in Cape Ann regularly. John Henry Twachtman, another Impressionist, started working in Gloucester at the turn of the century, staying at a fashionable hotel on Rocky Neck. Rocky Neck was a ramshackle little harborfront peninsula of shipyards and shacks. It had been an island at high tide, until 1840, when a causeway was built. The causeway made it possible to build a luxury hotel there, though it was down the cove from a paint factory and surrounded by decrepit waterfront shacks. But it also had Winslow Homer's view, looking out from the cove at

Ten Pound Island and a swath of Gloucester Harbor that turned radiant violet in thrilling buttery sunsets. There, Twachtman created soft and seductive paintings and taught classes, bringing in young artists who long after the hotel closed, still found Rocky Neck to provide an ideal and inexpensive place for artists to live and work.

Sunset over Ten Pound Island

chapter Seven

THIS GLOIRE OF GLOUCESTER

. . .

*I'm up this morning at dawn and my whole soul cries out
again looking out my door and seeing the early morning
sun . . . this rarest of all paintings of just this* gloire of
Gloucester *I can look out or walk out and find . . . and for
all pious people who think that history is problems, and that
we must now live, and with problems, Fitz Hugh Lane's
immaculate retention in paint of what is still my eye awakened this morning, out my door, on my blood, Saturday
morning walking it until I have lived on an ability and
plane of 48 hours of maximum production and human joy
(it is now Monday A.M. 7:45) I say shoo: this is, and can't
be bettered.*

—CHARLES OLSON,
LETTER TO THE *GLOUCESTER DAILY TIMES*, FEBRUARY 7, 1968

THE 1920S AND 1930S, THE PERIOD KNOWN AS THE GREAT
Depression, a time of notable hardship, is looked back on in
Gloucester with considerable nostalgia. Not only were the bad-
old-days good times for fishing, but the art movement which it
had drawn reached its height.

In the beginning of the twentieth century, led by Europe,
enormous changes were taking place in Western art. The American artists who visited the Armory Show in New York City in
1913, a show that introduced the work of van Gogh and other
post-Impressionists to America, left with their vision perma-

nently altered, and a few of these Americans took paintbrushes and their new vision to Gloucester. One such artist was an unknown painter from Nyack, just north of New York City, named Edward Hopper. He had been to Gloucester the year before the show, at which he sold his first painting, which was of a sailboat. It was an ironic beginning for the rare painter in Gloucester who almost never painted boats or the sea.

Ten years later, in the summer of 1923, Hopper returned to Gloucester, which was now full of New York painters, and while wandering looking for subjects, he met another New York artist, Jo Nivison, whom he later married.

He was a quiet man and he painted quiet, almost silent paintings, stark in their simplicity and powerful in their use of light and shadow. Unlike most of the artists who came to Gloucester, Hopper was not drawn to the maritime life, the light on the water, not even the rocks that would later enthrall Marsden Hartley and many others. "At Gloucester," Hopper wrote, "when everyone else would be painting ships and the waterfront, I'd just be going around looking at houses. It is a solid-looking town." He especially liked the sturdy nineteenth-century houses of Portuguese Hill, which he said "had the boldness of ships."

John Sloan, another major American artist whose vision was altered by the 1913 Armory Show, started spending his summers in East Gloucester the following year. He shared a house in front of the causeway entering Rocky Neck with other New York artists, including the twenty-two-year-old Stuart Davis. The first few summers Sloan produced nearly one hundred paintings each summer, including *Sunflowers, Rocky Neck*, a Gloucester homage to van Gogh. But he felt uncomfortable among the fishermen and rugged waterfront characters of Gloucester. After an altercation with a drunk he said that he would make "no more attempts in that fishermen's town to paint in the street."

In 1920 a local publication printed a poem called "Them Artists," which asked, "What is it sickens with disgust the Gloucester sailor man?" The answer, of course, was

It's these everlastin' artists, a setin' all around
A'paintin' everything we do from the top mast to the ground.

The local attitude was well summed up at the conclusion:

For they put us into picters and they think its just immense
They call it "picteresque," I b'lieve but it certain isn't sense.

When the fishing was at its height, from 1850 to the 1920s, so was the art. In East Gloucester farmers complained that artists' turpentine-soaked rags and discarded materials were making their cows sick, and some fishermen complained that the artists were always around them, getting underfoot. Boys climbed to rooftops to peer into studios where models posed naked.

But the people of Gloucester and especially Rocky Neck were getting used to the artists. Charles Movalli, a Gloucester-born art critic and painter remembered, "painting on a wharf and getting in the way of fishermen unloading. I apologized and a fisherman said, 'You have your job, we have ours.' " Most fishermen tolerated the artists and their easels on the piers and even aboard their vessels.

By the 1920s Gloucester had become almost as famous for its painting as it was for its fishing. Harold Bell remembered, "Rocky Neck was cheap with buildings ready to fall in the water. A bunch of shacks had one shower. Kids, wannabees, artists all rented cheap rooms there. It was fun to be part of that crowd."

In 1916 a gallery had been established in East Gloucester near Rocky Neck called the Gallery-on-the-Moors. The gallery's stated mission was to show "prominent American artists" who worked in Gloucester. The first show in September 1916 exhibited the works of forty-three painters and five sculptors. Eleven paintings were sold. The gallery started drawing New York critics, and works were selling for high prices although there was a tendency that has remained in the Cape Ann art world to exhibit mostly traditional maritime work. Avant-garde modernists such as Sloan and Davis were not popular, probably not even understood with their Gloucester scenes in flaming colors, in Davis's case with compositions suggestive of cubism. By 1919, Sloan had relocated to Santa Fe, though Davis continued to spend summers in Gloucester until 1934 and visited throughout the 1940s. Davis once wrote, "A great art requires a great audience of at least one person." In Gloucester he found a few more than that, though by his death in 1964 he had found a great many in other places.

GLOUCESTER THE FISHING TOWN was a town not only of painters but of writers. They, too, came in the summer. As a

boy, T. S. Eliot's family had a home on Eastern Point and starting in 1895, when Eliot was six, he spent every summer there
with his family until 1910 when he completed his education at
Harvard. His brother Tom continued in Gloucester but the
poet moved on to England, always remembering both the lore
and the perils of sailing off Cape Ann's rocky coast. Eliot was
convinced that William Eliot, of Salisbury, England, who
drowned in the famous 1635 shipwreck, for which Thacher Island was named, was his first American relative. Maritime
historian Samuel Eliot Morison knew the poet and tried, unsuccessfully, to convince him that this was not true.

In "The Wasteland," the 1922 poem that secured Eliot's
place as a master of modernism, a poem filled with improbably
fused autobiographical references, there is a section called
"Death by Water." The ten-line passage is the shortest part of
the poem. Originally it was much longer, with a haunting description of cod fishing the Banks in a Gloucester schooner.

Then came the fish at last. The eastern banks
had never known the codfish run so well.
So the men pulled nets and laughed, and thought
of home, and dollars, and the pleasant violin
at Marm Brown's joint and the gals and gin.

But, as often happens at sea, the laughter didn't last. Two
dories were lost and they went "scudding with the trysail
gone" and "leaping beneath invisible stars," and soon there is

an eerie song of white-haired women from the rigging and he is "frightened beyond fear, horrified past horror."

It might have been the most celebrated Gloucester story of all time had it not been done in by another Gloucester story. On his way home to London from Lausanne, Eliot stopped off in Paris to show the poem to Ezra Pound, who advised him to cut the passage, which Eliot did, along with many other revisions from the master, and Eliot dedicated the resulting poem to Pound.

Rudyard Kipling, at the height of his career, resided in the town center in the summer of 1894 while researching *Captains Courageous,* his novel of cod fishing the Grand Banks on Gloucester schooners. His only experience at sea that summer

Young T. S. Eliot on the family porch in Eastern Point in 1897
(HENRY WARE ELIOT COLLECTION, SAWYER FREE LIBRARY)

was a brief journey fishing pollack in which the author was too seasick even to take notes. Yet *Captains Courageous* is remarkable for its details—things Kipling never saw from his safe perch in Gloucester, such as the "ugly, sucking, dimpled water" of the Grand Banks.

Not all the summer residents were either artists or wealthy. By the twentieth century, working-class people also followed the artists and the wealthy to summer in Gloucester. Charles Olson's father was a postal worker in Worcester. Starting in 1915, when Charles was five years old, the Olsons spent family holidays in Gloucester. The mayor of Gloucester at that time, Homer Barrett, a small man who fidgeted with his cigar, was building inexpensive vacation cottages at Stage Fort Park, the original site of 1623 Gloucester. The Olsons eventually moved to a two-story cottage just out of the park, a little house with nautical ornaments including a pilothouse wheel on which was marked the cottage's name, Oceanwood.

Charles Olson grew up to be a huge man, six feet eight inches tall, even bigger than James Merry, Dogtown's ill-fated matador. When, after his sophomore year of college, he took a job as a mail carrier, the *Gloucester Times* announced, "Carrier Force Adds College Giant to Fernwood Route."

Shunning his father's advice—it was now the height of the Depression—to take advantage of the money that could be earned around the docks, Olson looked for that other Gloucester, the Bohemian artists' Gloucester. Every day he would row a boat across Gloucester Harbor to a theater group in Rocky Neck. It is typical of Gloucester's inherent contradiction that this working-class kid started a stellar literary career with the help of people with Gloucester connections. In Gloucester he met Edward Dahlberg, the leftist social-realism novelist who was celebrated for his depictions of the American working class. Dahlberg had come to Gloucester to escape hot summers in Greenwich Village. Through Dahlberg he met Marsden Hartley. He also corresponded with T. S. Eliot, a Gloucester connection that came about through Eliot's close tie to Ezra Pound. After World War II, when Pound was locked in a mental institution in Washington to escape conviction for treason because of his World War II support of fascists, Olson traveled to Washington to befriend the imprisoned poet. Through Olson's connections with Pound and Eliot, his first book, about the works of Herman Melville, was published.

Olson went on to become one of the important post-modern poets and, in fact, the man who coined the phrase "post-modern." When he died at the age of sixty in 1970, his burial in Gloucester was attended by Allen Ginsberg and other notables of the poetry world. One of the poets at the burial, Vincent Ferrini, an old friend from Gloucester, was in 1998 named by the city as its official Poet Laureate. That is Gloucester, a town built on a fishing industry but with its own Poet Laureate.

Ferrini was from Lynn, the son of an Italian immigrant cobbler, and first came to Gloucester in 1948. "I felt an explosion of my mind and said this is the place I want to be." Being Italian and working-class had much to do with his attraction to Gloucester. "The working man has so much to say and he has been relegated to a second-class position in life," he said in his small and modest ground-floor apartment in East Gloucester that was originally an artist friend's studio. In his nineties and still Poet Laureate, Ferrini looked like a cartoon of a snake, his long thin head with a rounded bald top, slit-like eyes, so slender that he seemed to have no shoulders or hips.

Asked if he had duties to perform in his capacity as Poet Laureate, his answer was simply, "No. They're afraid of me." And it was true. Official Gloucester did hesitate to call on him because no one could ever be sure what their Poet Laureate might say nor in many cases what he meant by it. He often said that the "transvaluation of values" was what most interested him, and when asked to explain, gave as an example, "Women will change America because the women have balls and the men don't."

Whenever discussions came up of Gloucester's problems, a frequent topic, he regularly asserted in a voice indicating that this was *the* answer, "The fish is in us," which seems to be a reference to his four-volume opus *Know Fish,* four of the forty-two books of poetry and plays that he published either by himself or with small presses between 1941 and 2004.

Other Massachusetts writers from blue-collar ethnic backgrounds have been drawn to Gloucester for the same reasons as Ferrini. Israel Horovitz, who became a New York theater sensation at the age of seventeen, came from Wakefield, Massachusetts, not far from Cape Ann. His father was a truck driver who at the age of fifty went to law school. The Horovitz family vacationed in Gloucester and liked it for its beauty and its working-class culture. In 1978 Israel Horovitz, now a recognized playwright, bought a home in Gloucester and the following year established the Gloucester Stage Company in East Gloucester, where he tried out all his new plays until resigning

from the group in 2007. Gloucester people are not easily impressed and speak their mind. When Horovitz went running in
the early morning, people would shout out their assessments of
his most recent play from passing cars. Recently a local ran
after him with a one-act play manuscript that he hoped
Horovitz might produce. Everything is possible in Gloucester.

ANOTHER GLOUCESTER ART COLONY developed in a little
bay called Folly Cove, by the quarries just before the Rockport
line. It was a picturesque spot where fishermen used to haul up
their dories, with a pier where sloops were tied up to be loaded
with granite from the quarries. Artists began moving to Folly
Cove after Charles Grafly (1862–1929), a widely recognized
sculptor and much-sought-after teacher, arrived there in 1903
and gave summer classes. Among the next generation who followed Grafly was Paul Manship (1885–1966), an early practi-
tioner of Art Deco, most known for his Prometheus fountain
at Rockefeller Center—which he never liked—and the even
younger Walker Hancock (1901–1998), who did monuments in
Philadelphia and Washington, D.C. "I fell in love with Folly
Cove, as it was in those days, very primitive and beautiful," Hancock recalled in a 1977 interview for the *Smithsonian Art Archives*.

In 1913, George Demetrios (1896–1974), a poor teenage
Greek immigrant from Boston, came to study with Grafly and
himself became a prominent sculptor. He married a dancer,
Virginia Lee Burton, whose mother was an English poet and
whose father was the first dean of the Massachusetts Institute

of Technology. In addition to dancing, she was interested in art and had gone to Folly Cove to study under Demetrios. She wrote and illustrated seven enormously successful children's books, including *Mike Mulligan and His Steam Shovel* (1939) and *The Little House* (1942), about their Gloucester house among the apple trees. She also established a design studio for linoleum block prints on fabrics. The group of forty-three mostly married women raising children, known as the Folly Cove Designers, became legendary for their prints for curtains, tablecloths, linens, wallpaper, clothing, and Christmas cards from 1941 until 1968, the year after Virginia Lee Burton died.

THERE SEEMED TO BE two distinct groups of artists in Gloucester: the New Yorkers and the maritime painters. The New Yorkers were more sophisticated. They were modernists pushing the possibilities of twentieth-century painting. The maritimers were more traditional painters of seascapes and maritime scenes. Some were of little note and others of them were masters, such as Gordon Grant from San Francisco whose maritime scenes painted and drawn with great accuracy and personal flair earned enduring reputations. Among Gloucester's maritime painters, Emile Gruppé was a dominant force and a notable character in a town that prides itself on colorful characters. His father, Charles Gruppé, a Canadian-born painter of considerable renown, moved to Cape Ann with Emile in 1925. They had lived for a long time in Holland, and Charles

Gruppé's land- and seascapes were clearly influenced by Dutch landscape painting and Impressionism. Emile was clearly influenced by Charles, whom he called "the great painter." But Emile, who grew up studying under his father and was almost thirty before they arrived in Cape Ann, was much more of an Impressionist and had a hotter, more vibrant palette.

Gruppé not only painted fishermen, he befriended them, and he also took up fishing. His painting lessons were popular events. Charles Movalli, who later wrote four books on painting with Gruppé, remembered his father's excitement about Gruppé demonstrations. Art had already become a way of life in working-class Gloucester. Movalli's father drove a heating oil delivery truck and his parents would gather blueberries and sell them in the street for extra money to buy paintings. Among the Cape Ann masters they bought was Gruppé. "My father loved Gruppé," said Movalli, "and my mother bought him ten Gruppé demonstrations for his birthday and I went with him. I saw him give demonstrations in the 1960s and thought somebody ought to do a book on this guy. He was a big bear. He seemed to have a huge head with a mane like a lion. He loved to talk about his sales—how much he was selling and for how much. It was a split personality—on one side very poetic, on the other a rough commercial character. You never knew which you were going to get. His demonstrations were just him painting and offering a few comments. It was 'paint along with Gruppé.' People loved it. There were a lot of Gruppé groupies."

Gruppé would tell his students not to copy him. Once a student asked, "Suppose there is a painter you admire so much you cannot help but imitate him?"

Gruppé's answer was "Then you have to learn how to hide it."

He would set up an easel along the Gloucester waterfront and start painting and before he was finished someone would buy it. "Just by chance," said Movalli, "his taste was exactly their taste." This was not the case with all Gloucester artists, especially the New Yorkers such as William Meyerowitz, a Russian immigrant who had built a reputation in New York before moving to Gloucester and was in some ways a bigger name than Gruppé. The locals did not understand Meyerowitz. But Gruppé did. He used to serve as the auctioneer at Gloucester auctions and would buy up all the Meyerowitzes he couldn't sell. Sometimes, when haggling over a painting, he would say to the collector, "Buy this painting and I'll throw in a small Meyerowitz."

Gruppé and many other Gloucester and Rockport artists paid bills with paintings. An East Gloucester liquor store acquired an exceptional collection, especially of Gruppé, who at times used to pass out from alcohol at his easel and would hire someone to get him home so the police wouldn't find him. He was forever going on the wagon and quitting smoking. A Rockport tailor, Bob Kline, amassed an impressive collection and so did Melissa Smith in her Rockport restaurant. The town of Rockport did well, too, and today there is a significant collection at the fire department, the town hall, and the hospital. But people like Movalli's father also bought artists when they were still affordable and it is not uncommon to see good and occasionally even great paintings in ordinary people's homes, shops, and offices in Gloucester and Rockport. As Samuel Hershey, a midwesterner who became a leading twentieth-century

Rockport artist, commented in a 1978 interview, "In any community other than this one, how many pictures do you see in people's houses?"

A LARGE NUMBER of painters settled in and painted Gloucester's split-off sister community, Rockport—the town, the harbor, and the quarries. One red fishing shack in the harbor was painted repeatedly by Aldro Hibbard (1886–1972), a native of Cape Cod who became one of Rockport's most influential painters, and by Dutch-born Anthony Thieme (1888–1954), who alone is said to have done four hundred paintings of the shack, and by their students and eventually by most everybody.

The author's rendition of Motif #1

According to legend, the artist Lester Hornby (1882–1956), known for his etchings and drawings, taught in Rockport in the summer and so many of his students presented images of this one shack that at one point he said, "What—Motif #1

Painting the harbor from Marilyn Swift's East Gloucester art class

again?" The scene has been called that ever since. Samuel Hershey once explained the attraction of Motif #1: "That thing hangs out there like a mother hen in the harbor. It dominates, or it did . . . And of course, it was red, let's not discount that."

In all, there have been hundreds of artists working on the tip of Cape Ann, at least eighty with prominent places in the history of American art. Artists still work here, including oil painters, watercolorists, and metal sculptors. Increasingly, they are home-bred. Chris Williams grew up in Rockport, the grandson of a Long Island sheet metal producer who moved there and opened a machine shop where his son worked and his grandson was trained to work with metal. At the age of twelve, Chris began working in the shop, sweeping floors. He learned about machinery—lathes and stamping and other shaping tools. He would spend hours in the Rockport dump after it was closed. He found bicycle parts. He made coasters. "It was always about reusing something. I once found a little three-horsepower motor and got it working. I loved watching it run, just sitting there on the table," said Williams.

Growing up where he did, the idea of artists was not unfamiliar. "I remember poking through the windows at painters working and talking with them, and I did things on paper with them. It put a seed in me somewhere," he said. But to him being an artist was something for "the trust fund babies and retired people."

"I had no paintings in my house growing up," Williams said, and then added, "a couple of inherited ones but no one went out and bought art." When he was thirteen he wanted to go to art school, but his parents opposed the idea. "There was no encouragement." But, he said, "If I had gone to art school I would have been less original."

He began by making Christmas presents. Then he started working on larger pieces, welding together scraps he found in junkyards. He made many of his own tools. Benders, squeezers, pushers. There were no names for such invented tools, so he started naming them so that when helpers worked with him he could ask for what he needed. Bob and Frank are two bending tools made from pipe that he always uses together and are named after an inseparable couple he knows.

Soon Williams established his own metal shop in the dilapidated backstreets of downtown Gloucester. He started making life-size animals with complex inner rods to produce a lifelike thick metal skin. First giraffes, then horses, a rhinoceros— huge pieces for which he was finding more and more commissions. In his hands metal seemed an organic substance.

Many people in Gloucester and in the art world believe that, as with fishing, the best days of Cape Ann art are behind them. Many artists complain that neither the art establishment nor the market permits them to be experimental. But that had always been true. Gloucester's art world began with Fitz Henry Lane and was always steeped in a tradition that many of its painters, such as Davis and Meyerowitz, never quite fit. The Gruppés—Emile and his father, Charles, but also his brother Karl, his sister Virginia, his son Robert, and Karl's son, Charles C.—and their paintings of the sea, the docks, the fishermen remain the mainstream of Gloucester painting.

Paul Ciaramitaro, whose family came from Terrasini like the other Ciaramitaros, had started sketching while working on Sicilian-owned fishing boats, and painted seascapes in oil. "I feel so lucky I was born in Gloucester," said Ciaramitaro, who studied under Charles Movalli. "But I got a lot of free advice from good artists." He attributed his becoming an artist to his Gloucester upbringing, but his heros were Picasso and Braque. He liked paintings where the image is broken up, planes are turned at angles. A man from Baltimore bought a number of his neo-Cubist pieces but he never sold another one. "I would do more, but you just can't sell them," he said. And so Ciaramitaro also concentrated on maritime painting.

The decline in the standing of Gloucester art posed few problems for Gloucester. In fact, it was barely noticed. "There probably isn't important work around," said Movalli of the hometown he loved, "but who cares. I think it's nice to have a lot of people around to talk painting with. Some of the least painters have wonderful things to say about painting. I'm not going to worry about it too much as long as I can talk to people and have fun."

But of far greater concern was the decline in the artists' subject, the fishery.

Chapter Eight

WHILE GLOUCESTER BURNS

· · ·

*The feeling that "something should and must be done"
has prevailed for some time.*

—*GLOUCESTER DAILY TIMES,* OCTOBER 3, 1911

ASKED ABOUT THE CRISIS THAT HAD STRUCK THE GLOUCES-
ter fishing industry, Gus Foote, a rugged-looking half-fisherman
half-politician, said, "Well, the fishing industry has been dead
many times." Foote's family comes from Newfoundland, the oldest and possibly the harshest fishery in North America. Foote, a
heavyset man with the rough-hewn charm of a big-city political
operator, was Gloucester-raised, his family vessel having gone
down with all hands on the day he stayed home to look after his
ailing father. Gloucester City Councilman for the downtown
waterfront district since 1975, he knew more than a little about
the ups and downs of fishing. But if he was not worried, he was
one of the few who wasn't.

It is true there had been troubles before. Books and articles
from the beginning of the twentieth century describe "the current fishing crisis" and discuss whether or not Gloucester had
a future.

Gloucester has been primarily a ground-fishing port. Gloucestermen fished for the bottom-feeding white-fleshed fish of the
continental shelf and the Banks. Cod, haddock, whiting, pollack,

and halibut had been their mainstay. But Gloucestermen also hunted the so-called mid-water, between the surface and the bottom—for darker, oilier "pelagic" fish, such as bluefish, herring, and mackerel and these fish, too, have been a traditional part of the Gloucester diet. Here is a Gloucester recipe from 1830.

SKULLY-JO FISH BAKE

Cut cleaned herring or mackerel fish into thick pieces. Place in a stone jar with 30 peppercorns and a half teaspoon of salt. Add one blade of mace and a bayleaf. Peel one shallot and add to fish. Pour in one gill of vinegar, and tie a brown paper tightly over the top of jar with a string. Put jar, with contents, in a very slow oven to bake 6 hours. The fish is to be eaten cold.

And there were always shellfish—lobster, shrimp, scallops, and clams—to be harvested, too.

MISS E. GROVER'S LOBSTER SALAD

Boil two or three eggs three minutes, mix them with the red and green of the lobster; add one and one-half teaspoons full mustard, one-half cup melted butter, one-half cup vinegar, a little salt. Heat it or not.

—GATHERED TREASURES: LADIES OF

THE INDEPENDENT CHRISTIAN

CHURCH, GLOUCESTER, 1884

The dory fishermen who chased cod and halibut baited their hundreds of hooks with herring, a migratory fish that will swim in one area for a decade or more and then suddenly disappear. Medieval northern Europeans, highly dependent on their herring fisheries, had an array of superstitions about what caused the herring to disappear. Sometimes it was interpreted as a punishment for adultery in the village. But in Gloucester, rather than worry about their marriages, the fishermen moved on to a different species. If one species was down, another was up. It seemed that there was always something to fish.

Gloucester fishermen were not particularly innovative about types of gear or types of vessels. This was in part because they were reluctant to spend money to upgrade, and partly because they didn't often feel that they needed to, because of less competition in their grounds.

It was in the North Sea that innovations took place because this was a body of water, rich in fish, that was surrounded by competing great European fishing nations. And it was in the North Sea that the crisis in fishing first began. For centuries, North Sea nations had been struggling to bring in ever larger catches, with little signs of any depletion of fish stocks. In the early seventeenth century, around the time Gloucester was founded, the Dutch had two thousand vessels in the North Sea fishing for herring. The English responded to the competition by banning foreign fishing vessels within fourteen miles of the British coastline, which was the distance visible from a masthead.

In England, dragging nets was an old idea. The British had

first used a "beam trawl"—a net suspended from a beam and dragged through the sea—in the fourteenth century, calling it "the wondrychoum." But the size of the nets on beam trawls was limited because a sailing vessel did not have the power to haul a huge net or a long beam. The beam gradually did get longer, stowed on the port side of the vessel, but a beam trawl under sail power was still very limited fishing gear. In a good hour it could sweep only two miles, and the vessel struggled to move faster than the tide so that the net would stay behind the beam. But in 1376 fishermen petitioned Parliament to ban it, or at least increase the size of the mesh, because it swept up fish indiscriminately, taking many undersized young fish. Though it was not banned, fishermen in Cornwall and Scotland continued to denounce this type of gear, insisting that their hook-and-line technique avoided the bruising of nets, thus harvesting a higher quality fish. But they also insisted that the beam trawl damaged spawning grounds and would decimate fish stocks. In the seventeenth century, when Gloucester was new, Scottish fishermen petitioned Charles I to protect their fishing from "the great destruction made of fish by a net or engine now called the Trawle."

But the potential of dragging a net through the water and hauling up everything in its path had obvious advantages over setting lines with baited hooks. In addition to requiring no bait, a beam trawl seemed certain to haul in a much higher percentage of the fish it passed. By 1774 beam trawling became one of the principal fishing techniques in the North Sea.

In the mid-nineteenth century, innovation was aimed at improving the quality of fish, getting it to market fresher. Well boats came into use—ships with a tank of seawater to carry to market a catch offloaded from a fishing boat—enabling fish to be landed fresh while fishermen remained at sea fishing for long periods of time. They needed to stay out to sea because once the quality of fish in England, and most notably in London, dramatically improved, demand for fish rapidly increased. In 1848, the port of Grimsby, on the North Sea at the mouth of the Humber, got a rail connection to London. Because it was a large port, capable of storing ice from Norway, it became the premier port for quality fish for London.

In the 1870s, eighty years after the first steam-powered vessels were launched for transportation, most fishermen were still using sail power. It was at about this time that the otter trawl was invented. This was a net, first used by the British in 1874, that had no beam but instead had "doors"—flat slabs of wood or iron on either side that caused the sides of the net to stay open. But the otter trawl worked only at a constant speed, requiring a more reliable and more powerful energy source than wind and sails. Steam power was needed for an efficient otter trawl.

At the time, steam was gradually making its way on board. In 1876 a fishing vessel was launched with a steam-powered capstan, the rotary device for hauling in and out the nets, and in 1881 a vessel was launched that used steam power as an auxiliary to sails for dragging nets. But even with the dual energy, it did not have enough power to drag nets efficiently in open seas or deep water. The southern part of the North Sea is an extremely shallow body of water, which is why it was so rich in fish, and this boat worked primarily on the even more shallow Doggers Bank, a shoal like Georges Bank where the shallow water was teeming with fish, albeit in waters not nearly as treacherous and unpredictable as Georges and the Grand Banks.

In 1881 the first vessel built for dragging under steam power, the *Zodiac*, was launched from Grimsby. Soon, steam-powered vessels were used to speed catches to Grimsby, and its rail line and port organized a system, used until 1901, that placed the entire fishing fleet under a single command, an admiral. The boats would stay fishing in the North Sea for as long as ten weeks at a time, all the while offloading their catch to the fleet's carriers, which sped the fish to Grimsby.

Over the next ten years four things happened: steam engines got more powerful, capable of dragging four times as deep as the sail-powered draggers; this opened up new grounds to dragging, including the deeper northern part of the North Sea and the waters around Iceland; Britain became the greatest fishing nation in the world in terms of tons of fish landed; and the fish stock of the North Sea started showing signs of depletion. But because the new boats could fish in places sailing ships couldn't, they were less dependent on the rapidly thinning traditional grounds. The whole world was opening to them.

In the North Sea, the drop-off in catches after ten years of dragging was so dramatic that even Thomas Huxley, the famous Darwinian scientist who had repeatedly assured the world that overfishing was a scientific impossibility, recanted his celebrated opinion in his old age after seeing the destructive power of draggers. From the late 1870s on, the British regularly convened commissions aimed at curbing the destruction by trawlers, but the size, capacity, and numbers of such vessels meanwhile were increasing at a steady pace.

In the 1880s a sail-powered beam trawler came over to Gloucester. Having their first glimpse of beam trawling, Gloucester fishermen thought it was a bad idea, that it took too many fish and would damage the ocean floor. Nevertheless, in the 1890s Gloucester got its first beam trawler, which had an auxiliary steam engine to supplement the sail power.

By 1911 New England fishermen were uniting with those of other regions to demand that Congress ban the new practices. The *Gloucester Daily Times* of October 3, 1911, reported:

> The inception of what is believed will result in a strong and mighty protest of the fishing and vessel owning interests of the Atlantic and Pacific coasts against the use of the otter and beam trawl took place at the rooms of the Master Mariners Association yesterday afternoon, when at one of the most largely attended meetings in recent years, the association went on record as condemning as strongly as possible this method of fishing and appointing a committee to take up active work of crystalizing the feeling among master mariners, fishermen and fishing vessel owners at any and every possible port and later urge governmental action in the shape of prohibitive legislation.

Government never acted, but it is clear from this Gloucester newspaper article that the ensuing tragedy of the next hundred years was plainly seen back in 1911. The article stated that there was a feeling, not only in Gloucester but also in other New England ports, that "something should and must be done" and that "the continued operation of these trawlers scraping over the fishing grounds and destroying countless numbers of young and immature fish, is the greatest menace to the future of the fisheries, the greatest danger the fisheries have ever faced along this coast."

The greatest danger. It was all understood a century ago. Gloucester fishermen had simply to look at Europe and especially Great Britain to see the future. According to this article, beam trawling had increased the fishing capacity in the North Sea to 14,000 times the capacity of the former sailing fleet. The article claimed the "wasteful destruction of immature fish" was one of the primary problems and that this destruction "cannot be obviated by regulating the size of the mesh nor by returning the undersized fish caught to the sea." The article concluded that, as American and European regulators have only recently come to understand, "the only feasible method is to close off fishing grounds or prohibit the landing of fish." It cited a moratorium that had recently been declared by the Scottish fishery Board in the Moray Firth.

The *Gloucester Daily Times* article further asserted that the history of the British fisheries showed that if New England did not ban trawling in its infancy, the trawler sector would later become too powerful to stop. After devoting a considerable portion of the newspaper to describing the gradual destruction of British fisheries and the failure of government to stop it, the article concluded, "While Nero fiddles, Rome burns."

BY THE 1930s, British engine-powered vessels were traveling ever farther to find fish, and the British government was
discussing an endemic problem in their fishing fleet causing
the disappearance of fish stocks. But incredibly, in Gloucester

Wooden-hulled side trawlers in Gloucester Harbor, circa 1925
(PHOTO BY ALICE M. CURTIS, COURTESY OF BODIN HISTORIC PHOTO)

there were fishermen still working under sail power until 1960, their schooners still made in neighboring Essex well into the twentieth century.

Following the Civil War, more and more of the limited solid land by the marshes of Essex was taken up by shipyards. In time, fifteen or more different yards were operating there. They built yachts and steamboats and other vessels, but they

mostly built fishing schooners. In the mid-nineteenth century, forty or fifty of them were being turned out a year, ninety feet or more in length. A schooner hull was completed in two months and could then be sparred, rigged, and outfitted in two weeks. A fishing schooner usually paid for itself in two good years.

A captain who wanted a schooner needed to advance the builder only the cost of the lumber. The shipyard workers were paid months later, when the vessel was completed and the captain had paid up. Local stores gave them credit until then. The builders worked ten-hour days and the top salary was $2 an hour. Toward the end of the century, the shipbuilders went on strike and got the workday reduced to nine hours. A 1910 strike got it down to an eight-hour day. But the sooner they finished the vessel, the sooner they were paid. They worked six-day weeks.

From the causeway leading to the center of town anytime of year, unfinished new vessels mounted on stocks could be seen in all directions, but by the time of the Depression, only two yards were left and the industry slowly died. The last fishing schooner was launched from Essex in 1949. Gloucester fishermen finally turned to draggers, not because they could catch more fish but because engine power was less dangerous than sail power. Once the engines became available, fishermen were increasingly unwilling to work on a vessel that might get blown over in a gale.

One of the last Gloucester schooners to fish, the *Thomas S. Gorton,* owned by the famous Gloucester seafood company, was built in 1905 and remained in service until 1956. But that was not the end for the old wooden hulls of Essex. Without sails and with an engine installed belowdecks, these vessels continued to work the fishing grounds until late into the twentieth century. The net was dragged from the side, which took considerable skill, especially when hauling in a full net, because there was a risk of fouling in the propeller and destroying the net. Nets, especially before nylon monofilament became available in the mid-twentieth century, were extremely expensive—worth thousands of dollars. Damaging or losing one was a calamity. So the side trawler skipper would carefully circle about into the wind before hauling. Families, especially Sicilian families, would pass down their wooden side trawlers from one generation to the next, because buying a new steel-hulled stern dragger, or as they became known in New England, a bottom dragger, was too expensive.

For those who could afford it, equipment got better and better. The Gloucester fishermen who bought new boats bought steel-hulled stern trawlers, almost always on credit. Stern trawlers had been developed in the Pacific by installing more powerful engines that were capable of dragging a huge net off the stern of the ship. Fishing had switched from passive gear, devices that waited for fish to come by, to active gear, devices that chased down fish.

Stern trawlers also had the advantage that they had a lot of open deck space. The net was rolled and unrolled from a spool on the stern, the pilothouse was placed forward, and midship was open for fish processing. Because they were ever bigger vessels with larger nets pulled by larger engines, the vessels had larger hulls, which meant more space below for storing fish. These vessels could travel farther, stay out longer, and catch more fish; eventually the largest ones became known as "factory trawlers," seagoing factories for catching and processing fish.

But stern trawlers were already an old idea by the 1970s when they started coming to Gloucester. Gloucester had never been a major port for long-distance factory trawlers, because its fishermen had too many good fishing grounds near home port. Unlike factory trawlers designed to stay at sea for months, twelve days out was a long trip for a Gloucester boat; often a trip lasted only a week. The fishermen would take about thirty trips a year, which meant about three hundred days of fishing a year.

The new trawlers used "ticklers," chains on the bottom of the net that had a lot of movement and drove fish into the net, and "rock hoppers," bouncy rubber rollers that could hop over and between rocks. Bottom draggers could go anywhere. With rock hoppers and monofilament nets they no longer had to shy away from the rough, jagged bottoms for fear of damaging their nets. They could hunt down fish as they hid between rocks. Trawlers became the fishing equivalent of strip mining, destroying great swaths of ocean bottom as they dragged along.

After World War II, the technology developed for chasing submarines, such as sonar and spotter aircraft, was added to commercial fishing to efficiently locate schools and then scoop them up with nets. Joe Santapaola had one of the first sonars in Gloucester. He remembered thinking it was painfully complicated to operate. "It would make you sick," he said. "Take the eyes out of your head." But the equipment grew more efficient, and fishermen grew more skilled at using it. Gorton's praised the marvel of the new technology and boasted in their advertisements that fishing was no longer "hit or miss." But fishermen could see a problem. "With rock hoppers poor little fish don't have a place to hide," said Rich Arnold, a retired Gloucester fisherman.

The Novellos were one of those huge Gloucester fishing families from Terrasini. Sam Novello, the son of Lena, who founded the Fishermen's Wives Association, recalled eighty relatives who worked on Gloucester vessels. "I am the last," he said sadly.

Though the Novellos worked in the heyday of big-time trawling, they were not running factory trawlers that raped the ocean. For a long time his immediate family fished on his father's 1936 wooden-hulled side trawler named after his grandmother, The *Vincie N*, which today is a museum piece in the downtown waterfront. The *Vincie N* was considered innovative because it was designed for both mid-water and bottom dragging. The Novellos prided themselves on being creative modern fishermen, which in part meant implementing conservation ideas. Sam had used wider mesh in his nets a decade before it was required. "I wanted the other fish to live," he said.

For a time there was a fishing boom, but fishing got so bad in the 1950s that Sam Novello's father took the *Vincie N* shrimping in the Gulf of Mexico. If the catches or the marketplace made a traditional fish, such as cod, unprofitable, there was always the possibility of switching species. Dragging does not target a species the way hook-and-line fishing does. It scoops up everything in its path. A certain degree of targeting is done by determining the area that is fished and the depth at which the net is dragged. But a dragger net will still haul up a wide variety of other species, known as by-catch. Half of the fish in the net can be by-catch. Most fish hauled up in a net are dead from the rapid change in pressure or in such bad condition that they cannot survive, so the unwanted fish were simply dumped back into the sea as garbage.

But what was unwanted changed all the time. In the 1930s Gloucester had a diminished and backward fleet of about 150 vessels, mostly wooden-hulled, and the fleet was finding it hard to compete with foreign suppliers of ground fish. But in 1921 the filleting machine came to New England, and after it was combined with Birdseye's fast freezing process, fishsticks became an important Gloucester product, especially from Gorton's. Fishsticks were made of an unspecified white-fleshed ground fish. As fish stocks declined, Gloucester fishsticks

evolved quickly from cod to haddock, and then Gorton's
turned to redfish, a relatively slow-growing ocean perch that
fisherman had been dumping over the side.

For a time, fishsticks altered the Gloucester fishing industry. Filleting created an enormous amount of scrap—skin and
bones. The city began to complain about this scrap being
dumped in the harbor, and finally an ordinance was passed
against it. The fishing industry started promoting the scraps
as fertilizer, and one of their customers was a farmer named
John S. Rogers, who noticed that his boots were getting stuck
to the floor after working with fish skins. Rogers was an ama-
teur Classicist, and he remembered that the ancient Greeks had
made glue from fish skins, calling it *icthyokolla*. The process
had been lost. He experimented in his barn, as did a competitor
named Stanwood, buying all the skins left from the filleting

Bottom draggers in Gloucester's inner harbor

machines. While these two fought each other in court for years, they were outmaneuvered by a salesman named William LePage, who had learned a better glue process at Harvard and formed the Wm N. LePage Cement Company, whose glue became a national product and an economically important ancillary industry for the Gloucester fishery.

There was no shortage of redfish skin. In 1951 redfish for filleting accounted for 70 percent of all fish landed in Gloucester. But harvesting more than 150 million pounds annually of a slow-growing fish soon took its toll, and by the end of the 1950s there were no longer enough redfish to supply Gorton's. For a while whiting, a small fish in the cod family that had also been a by-catch, took the place of redfish, but the whiting supply was not dependable. By the 1960s a decline in ground fish was apparent. The fishermen of Gloucester, and of every other fishing port in New England, blamed the problem on "the foreigners." This was more than chauvinism, though that played a role as well. Giant factory trawlers from Japan and the Soviet Union, with government budgets supporting them, were sweeping traditional Gloucester waters such as Georges Bank. Gorton's and other fishstick makers started buying inexpensive blocks of foreign-caught ground fish, much of it caught in New England waters by giant trawlers that undersold New England cod at Gorton's. Rich Arnold said, "The problem all started with foreign ships. They would move into an area where you caught plenty of fish. They would fish there for three days with their huge nets and no one could find any more fish."

Today Gorton's still announces itself with a very large sign in downtown Gloucester, its logo the fisherman that is also the symbol of Gloucester. But in truth "Gorton's of Gloucester," as the company likes to call itself, has been passed from one foreign multinational to another in recent years. It no longer buys any fish from Gloucester and its fishsticks are made with another fish in the cod family, frozen Pacific pollack.

chapter Nine

THE FISH IS IN US

. . .

Florence: Who knows anything about fish?
I mean fish don't tell you anything much, do they?
You ask 'em a question, they flop around . . .

—ISRAEL HOROVITZ,
NORTH SHORE FISH, WORLD PREMIERE AT THE
GLOUCESTER STAGE COMPANY,
AUGUST 24, 1986, NEW YORK CITY PREMIERE
AT WPA THEATER, JANUARY 12, 1987

SOMETHING HAD TO BE DONE.

New Englanders were not the only ones growing angry about foreign trawlers. As the vessels got larger and more menacing, and developed the capacity to go anywhere, it seemed most everyone was getting tired of "the foreigners" scooping up their fish. The Icelanders and the Norwegians were the first to propose exclusive national fishing zones. This was a reversal of international maritime law, which had always emphasized free access. In fact, denying access was often considered an act of war.

Iceland, long a backward, underdeveloped colony of Denmark, gained its independence in 1944 and wanted to base a new modern economy on fishing. But Icelandic waters had been a traditional British fishing ground, especially for cod and haddock, the first the favorite fish of the south of England and the latter the favorite of the north and the Scots. The British had been fishing there since at least the fifteenth century. But once their stern trawlers started depleting the North Sea, British North Sea ports, especially Hull, became long-distance ports targeting Icelandic waters.

In 1950, when Iceland announced a 4-mile limit, the British, who had been claiming and enforcing exclusive fishing zones in the North Sea for centuries, were suddenly outraged by the concept. Eight years later, when Iceland extended the zone to twelve miles, the British were ready to go to war. There were three shooting wars—cod wars—in 1958, 1971–73, and 1975, all dangerous skirmishes between the British Royal Navy and the Icelandic Coast Guard. The Icelanders kept expanding their zone and the British kept resisting. But all the while, in its negotiations for entry into the European Common Market, Britain argued for its own zone of exclusion to keep out its European partners.

In truth, what was happening was that fishermen were gaining the ability to catch more fish than were available in the sea. During World War II, for six years, there had been almost no fishing in northern Europe. As a result, the fish stocks built up to levels that have never been seen since. But by the 1950s, catches around Iceland, in the North Sea, and even off Cornwall and around Ireland, the Irish box, were noticeably declining. There were two contradictory responses to this. Everyone wanted the foreigners out, but also everyone wanted to build up their own fishing capacity. The foreigners were not being thrown out to save the fish but, rather, so that the locals could catch more. As the local fleet caught more fish, fewer fish were available. The solution was to have more vessels that could travel farther with more power and larger nets. The problem with these responses, it can easily be seen now, but as was clear to only a few then, is that this creates a downward spiral. The new fleet would further deplete the stocks and the catch would be less, and so the fleet would forever have to be expanded to continue catching the same amount, resulting eventually in fewer fish and, finally, no fish at all.

In the 1970s, following Iceland's decision to declare a 200-mile exclusive fishing zone, most of the seagoing nations of the world did the same. For the United States this was not a particularly difficult decision. In the country's early days, when it was little more than a dozen states concentrated along the Atlantic, and Massachusetts politicians such as Massachusetts native John Adams, a tireless supporter of New England fisheries, were leading figures, fishery issues were central to U.S. policy. But in modern times, such issues have been of concern to only a handful of the fifty states, and even Massachusetts leaders have rarely seen them as a priority. Of greater interest to Washington, a zone of exclusion had implications for *mineral* rights, for *oil* rights. In fact, to protect these rights, the United States had been fairly early in claiming its continental shelf, which by chance also includes the richest fishing grounds. By 1945 it was clear that parts of these shelves held valuable oil deposits, and then-president Harry Truman declared the United States the exclusive owner of the waters of its continental shelf, though the United States had not restricted access by foreign fishermen.

In 1976 the Magnuson Fisheries Conservation and Management Act went into effect in the United States. This legislation established the exclusive fishing rights to a zone of two hundred miles from all American coastlines and also a system of fishery management broken into eight zones, one of which is New England.

But managing fisheries, while not a new concept, was something with which no government had much experience. For centuries there had been fights over fishing rights—for instance, between the Dutch and English over North Sea herring—and there had always been periodic outcries to ban certain practices, such as when the Gloucester dorymen protested draggers. But there had never been a systematic government regulation of all activities of every fishing operation. Fishermen had been free to do as they liked, and they saw it as their challenge to catch as many fish as possible.

There were two underlying beliefs supporting this approach. One was that the marketplace would curb the excesses of fishermen, such as in the 1830s when Gloucester fishermen landed so many halibut that the price collapsed, ending for a time the halibut fishery. The other was the widespread belief that persisted even after Huxley changed his mind, that overfishing was biologically impossible. That nature is unconquerably resilient is in itself one of the more resilient and also more dangerous beliefs.

So while Iceland and Norway, after establishing their 200-mile zones, set up rigorous fishery management programs, the concept of fishery management in what would become the European Union, Canada, the United States, and most of the other nations of the world was that once their waters were rid of foreign factory trawlers, they could build up their own capacity to catch the fish that would now be theirs alone. In most countries the 200-mile limit actually led to increased fishing. The Canadian government built up a huge ground fishing fleet for the Grand Banks in Newfoundland, seeing it as an opportunity to provide jobs. In the United States, government support and subsidies were more difficult to obtain, but through easy money from banks and venture capitalists, including soft loans and loan guarantees, the capacity of the fleet in Gloucester was greatly increased. The predominant attitude was clearly expressed by Lena Novello in a 1978 interview when she said, as did many others, "There are plenty of fish out there now that the foreigners are away."

The first few years after the foreigners were ejected was a boom time for New England and Maritime Canadian fishing. This, of course, encouraged government and fishing companies to enlarge the fleet even more. And with the boom came new technology. This was when light monofilament netting came on the market, as well as diesel engines that were both smaller in size and much more powerful. The more powerful engines and lighter netting meant that much larger nets could be used. Fishermen learned to rig the doors—the two iron or wood-and-iron slabs that force the nets open when running in the water—farther out on long cables that scraped the bottom, helping to chase ground fish from their hiding places and into the net, but also doing even more damage to the seabed. This was also the age when space exploration was opening up still newer technology, including better navigation equipment that could be used not only to find the fishing grounds but also to find the fish with pinpoint accuracy.

Rich Arnold said, "In the old days all you needed for equipment was a watch and a compass, maybe a depth sounder and you went out and found the buoy. There were foggy days when you couldn't find it. Now all you do is punch a couple of buttons. You don't even touch the wheel. Just stay in the pilothouse and try to keep awake for an emergency."

Sam Novello, the last Novello to go to sea from Gloucester, said, "The technology ruined it. It made fishermen too on-the-money. Fish don't have a chance."

Fish stocks declined all over the world in the 1980s. But this was not always apparent. In the Canadian Grand Banks, where hindsight has shown that one of the world's most plentiful and famous fish stock, the northern cod, was being exterminated at an alarming rate, all warnings were dismissed because the catches were unusually large. Based on these catches, Canadian government scientists were estimating a huge stock size, and the government, in keeping with its anti-unemployment program, was pushing for more fishing. The inshore fishermen, a traditional fishery hand-lining or using rope traps hauled in small skiffs, were seeing ever fewer cod, and they protested that the breeding stock was being wiped out by the huge trawlers on the Banks. A few independent scientists agreed. But the Canadian government, backed by its scientists, not surprisingly chose the politically preferable position of increasing jobs and delivering good news over decreasing jobs and announcing bad news.

By 1992 it became evident that only about 10 percent of the cod population remained and by 1994 the Canadian government closed the Grand Banks to cod fishing. No more cod on the Canadian Grand Banks? At last the world started to understand. Overfishing was not only possible, it was happening. Huge, historic fish stocks could be destroyed within a few years by modern technology.

IN 1943 THE BRILLIANT British North Sea fishery scientist Michael Graham had warned, "Fisheries that are unlimited become unprofitable." He also stated, "Free fishing failed everywhere," by which he meant that there was no unregulated fishery in the world that was not faced with disaster. After the 1994 moratorium on the Grand Banks, fishery management became much more "earnest," though earnest has often meant merely more widely discussed, more covered by the press. The New England fisheries have been managed by a variety of approaches. First, there is the quota system. Each vessel is allowed to take a certain number of pounds of a species. The scientists estimate the size of each fish stock and the fishery management officials decide what percentage of that total— less than half, often a quarter—can be taken without threatening the stock's ability to reproduce.

Sam Novello still remembers his first encounter with the bureaucracy of fishing quotas. He was told that he was allowed to take eight hundred pounds of cod on each trip. "I caught 2,400 pounds. I came in and said, 'All right, I have fished three days' quota.' 'No,' they said, 'it's one day and you have to dump 1,600 pounds overboard.' That's crazy. Why can't I count it for three days? But they said no."

Quotas result in millions of pounds of dead fish being dumped overboard every year. Fishermen haul in the net. They call into the market on their cell phone—in the case of Gloucester, they call the Display Fish Auction on the downtown waterfront. If the price of cod is low that day and the price of haddock is good, they will keep the haddock and dump the cod. Why use up your quota of a species on a day when the price is low? Even if prices are good, once the fishermen are over their quota they are required to dump the rest.

There is also the issue of by-catch. In New England the approach has generally been to permit by-catch. A fisherman targeting flounder is permitted to land whatever cod turn up in the net. When the by-catch starts getting larger than the target catch, questions have to be asked about which is really the target. But even when the by-catch is considerably fewer than the target catch, it can still be significant. In 2007, for the first time a quota was put on a by-catch. The mid-water herring trawlers were landing so much haddock, a fish that swims in the upper range of the bottom, that a haddock by-catch quota was established. But this may have been motivated by concerns not only about by-catch but also about "stopping the foreigners," since the much-resented large, modern, mid-water herring trawlers that tie up in Gloucester Harbor are Irish-owned. It is the registry of the vessel, not the nationality of the owners, that determines where it has the right to fish.

One of the most controversial measures—the one most

hated by fishermen—regulates the amount of time that fishermen are allowed to fish, which in New England is done by the number of "days at sea" per vessel. The number of days allowed varies depending on factors such as gear type. Because the days at sea were determined as a percentage of the days a vessel had been fishing when the regulation was first imposed, it favors the most destructive kind of fishing, the large bottom draggers. Since these vessels could stay out at sea for weeks and every fifteen hours was counted as a day, they got the most days at sea. Many Gloucester vessels have slipped below fifty days of work a year, which means a fisherman needs several boats. Then after he uses up his days on one, he can come in and go out on another boat. When Iceland first got its 200-mile zone, it used days-at-sea limits, but soon discontinued the practice because they found it wasteful to encourage fishermen to have a huge underutilized fleet just to have enough days at sea.

One approach that has yielded some results is closing off fishing grounds for a certain period of time and then reopening them and closing other ones. The problem with this is that for it to work, nearby grounds such as the Gulf of Maine must be periodically closed as well as farther offshore grounds such as Georges Bank. But Gloucester fishermen have largely abandoned the long-distance grounds for the nearby ones because they do not want to use up days at sea traveling to distant grounds.

There is another way to regulate fisheries. Gear can be regulated. This is done in modest ways, imposing large mesh sizes in nets, restricting engine power, banning rock hoppers. But

from the historical perspective, stern dragging itself is the cul-
prit. That is what has endangered two-thirds of the world's
fish stocks. Would the solution not be to ban bottom dragging?
On the West Coast, regulators have had some success in banning dragging in large patches of the ocean. There again,
xenophobia helped because most of the largest draggers were
foreign-owned.

Banning dragging is a politically explosive issue because
the most powerful fishermen, the large-scale fishermen, are
bottom draggers. Massachusetts politicians believe they need
the fishermen to carry Cape Ann, and they need Cape Ann to
carry the North Shore, and they need the North Shore to carry
Massachusetts. They also think they need New Bedford, which
is a dragger port. This is largely political mythology, as is the
belief that the Cuban vote is needed to carry Florida. Nothing
brings out the superstitious side of politicians like a swing
voter, and Cape Ann, in Democratic Massachusetts, alternates
its support between Democrats and Republicans. In the 2006
gubernatorial race, the Republican Kerry Healey worked hard
for the fishermen's vote in Gloucester. Fishermen and fishermen's organizations backed her. Gus Foote, who happened to
be a Republican anyway, said, "She is great for fishing. Great
for Gloucester," and suggested that her carved likeness adorn
the bow of fishing vessels "like the old days." The incongruous
image of a carved likeness of the thin and proper politician
voluptuously ornamenting the rusted hulls of Gloucester fishing boats somehow captures the absurd contradictions inherent in contemporary fishing.

Healey's opponent, Deval Patrick, who was forced to apologize for appearing to associate Gloucester fishermen with drugs and alcohol, easily carried Cape Ann and even Gloucester without the support of fishermen. Most politicians believe that while they may not need the votes of fishermen, they do need to appear to be supportive of fishermen in order to carry the broader community. Once elected, Patrick proved to be supportive of fishermen as well, saying that they were so hard-hit by regulations that they needed federal economic assistance. Unfortunately, most politicians do not show their support for fishermen with innovative proposals that would risk angry denunciations, but by calling for the status quo. A bold proposal, such as banning bottom dragging, would be viewed as bad politics since it would infuriate the most powerful and vociferous sector of the fishing industry.

Mark Murray-Brown is an English marine biologist who is a regional fisheries manager specializing in bluefin tuna for the Gloucester office of the National Oceanic and Atmospheric Administration (NOAA), the chief federal regulatory body. "New England uses more effort control than quotas," he said. "You have to ask what is the social environment willing to accept. For New Englanders effort control is more acceptable. But it requires far more regulations. Regulating by effort control such as days at sea is very complicated."

But he recognized that the political will is not there to do the most significant measure—a banning of net dragging. "I'd like to see a forest of masts," he said jokingly. "The problem with regulating by eliminating draggers or others is that it means some fishermen will have their boats banned, and this becomes a very volatile issue."

No one is suggesting going back to schooners, though there is research being done in energy-saving auxiliary sails, and there is a movement to return to baited hooks. In Chatham, one of Cape Cod's stateliest and most expensive towns, where real estate priced out fishermen and other blue-collar families years ago, there improbably remains a small fishing fleet on the elbow of the Cape. There is also an organization called the Cape Cod Commercial Hook Fishermen's Association, which goes by the unpronounceable acronym, CCCHFA. The organization looks for innovative solutions to the fishing crisis. This in itself is an important concept because the great weakness of the numerous Gloucester organizations for fishermen is that, while they decry and denounce, they have few workable counterproposals. This makes them appear as though they simply want to overfish, which is not the case. Fishermen in almost every region have been the first ones to speak out against overfishing, which they see as a real danger that must be confronted by management. Vito Giacalone, an outspoken Gloucester fishing advocate from a large Sicilian fishing family, said, "The modern fisherman does not want to overfish. He can't stand throwing out fish. He recognizes the benefits of closing areas. He doesn't want to use small mesh and catch a lot of little fish."

But in Chatham, fishermen have gone further. The founding concept of the hook fishermen's association was to reject dragging. Its members fish only with hook and line and take care to land their fish, primarily cod, fresh and not bruised and banged up, as fish are in what is called the cod end of a dragger net. They gave their line-caught cod a marketing name, Chatham Cod, and it commands unusually high prices.

Cheap fish is part of the problem. Expensive fish is part of the solution. It is hard on the consumer, but why shouldn't people pay as much for a wild fish that someone risked his life to catch as they pay for wild meat? If fishermen must earn a living while catching fewer fish, they will have to be paid more money for each fish they land. The marketing of Chatham Cod has been a success, even in Gloucester. Butch Maniscalco, who ran early morning auctions at the Gloucester Display Fish Auction, said, "Chatham Cod is pretty much the best cod available and that is reflected in the price." The auction is a privately owned company on the Gloucester waterfront that handles between 85 and 90 percent of all the fish landed in Gloucester, but also sells fish from other parts of New England. Operations like the auction in Gloucester, which first opened in 1997, are critical to this new concept. Before there were such display auctions, in which the fish are put out in bins for bidders to view, poke, and examine, a badly handled fish brought the same price as one in perfect condition. Now fishermen have an incentive to handle fish carefully and land them quickly.

The Cape Cod hook fishermen had what environmentalists call "a sustainable fishery"—a fishery that catches fish in a way that ensures the fish stock population will be maintained. In 2006 they were exempted from the New England Fishery Management Council's tough ground-fishing regulations because the fishing gear they used assured that they could not overfish.

Here was an idea that seemed useful for Gloucester and many other places. Gloucestermen could sell off their bottom draggers—they might even be able to get the government to buy them in an effort to reduce the fishing capacity of the fleet—and they could line-catch cod or haddock, call it a Gloucester Cod, get it to the auction house fast, and demand a high price. Wouldn't fashionable New York and Boston restaurants pay handsomely for a high-quality fish that had the name Gloucester Cod?

But Gloucester fishermen have not been drawn to the idea, Vito Giacalone said. "The problem with the Cape Cod Commercial Hook Fishermen is that they are only targeting one species. It would not work for a multispecies fishery. And it is also difficult to predict. You can agree to use only a certain number of hooks on the line, but then you will hit a time when you need to change that."

But there are other problems as well. Cape Anners look down on Cape Codders, especially in the world of fishermen. There is a natural competition—it existed even in the 1600s— between the peninsula south of Boston and the one north of Boston. They share many of the same fishing grounds, including Stellwagen Bank, and some traditional fishing ports such as Provincetown are by boat very close to Gloucester, although Chatham is actually on the other side of Cape Cod. But to Gloucester people Cape Cod is not a "real" place and Cape Cod fishermen are not "real" fishermen. Real fishermen don't use hook-and-line.

There is also a class issue. Gloucester is so resolutely blue-collar that it sometimes becomes an obstacle. The leader of the Chatham group, Paul Parker, went to sea and worked as a fisherman but he was a graduate of Cornell with an additional master's degree, another reason he was not a "real" fisherman. Even though Gloucester is full of college graduates, even many from the Ivy League, especially nearby Harvard, Gloucester people love to snub such things. The town thrilled to a story in The *Gloucester Daily Times* about a Harvard professor who was caught stealing manure, because it showed what Harvard professors "have a lot of."

So if Parker at the Cape Cod Hook Fishermen's Association was not a real fisherman, according to Gloucester fishermen, what was he? He was an *environmentalist*!

There was a moment in 1979 when fishermen and environmentalists were allies, fought vigorously side by side and succeeded in blocking an attempt to drill Georges Bank for offshore oil. It was when Angela Sanfilippo took over the Fishermen's Wives Association and turned it into an important political force for the interests of fishermen. They succeeded and because of their victory, the Fishermen's Wives Association stayed in the fray, especially Angela, who exhibited the feigned toughness of a wrestling coach.

But while both have the goal of preserving fish stocks, there have been few alliances between fishermen and environmentalists since. In 2006, when the Gloucester fishermen and then-Mayor John Bell tried to block energy companies from building an offshore liquified natural gas terminal, an LNG, support from environmental groups was far less than it had been during the oil drilling debate. Natural gas, unlike oil, is a clean alternative energy. The fishermen tried raising environmental issues. Sanfilippo said it might explode. But the real objection, a very understandable one, was not environmental degradation but the fact that the gas terminal would require closing off an area just off Gloucester from commercial fishing. Limiting days at sea had made the nearby fishing grounds particularly valuable, and now they would be losing them to the Massachusetts energy plan. The fishermen felt that the environmentalists, who had lobbied to have so much taken away from fishermen, owed it to them to keep the energy lobby from taking away their closest fishery.

It seems to fishermen, not completely without reason, that environmentalists are insensitive to their sacrifice. The fishermen accepted the need for fishery management and submitted to endless waves of regulations and restrictions on their ability to earn a living. But environmentalists have often characterized fishermen in ways that are unfair and, to them, deeply insulting. British environmental journalist Charles Clover, in his book on global fishing titled *The End of the Line*, is in places understanding of their plight. But he also writes, "We have an outdated image of fishermen as rugged, principled adventurers, not as overseers in a slaughterhouse for wild animals."

Fishermen have tried hard to avoid a slaughterhouse, and they have done all the sacrificing. What have environmentalists sacrificed to save the seas? From the fishermen's point of view they work a hard and dangerous job, risking their lives to feed society, while environmentalists and regulators are safely on land, earning comfortable salaries off of the fishermen's problems. In 2000, a consortium of environmental groups sued the federal government for not being tough enough in their fishery management. They said that the federal regulators were not fulfilling their mandate to protect fish stocks.

Gloucester fishermen responded by demonstrating in the harbor with large signs that read, "Out of work and hungry? Eat an environmentalist."

The more fishermen suffer, the more money environmentalists raise for themselves. When the question is asked, "What can an out-of-work fisherman do?" fishermen have been known to reply, "Become a regulator; that is the growth industry." And some do. In 2006 the city of Gloucester fought hard to have the regional office of the federal fishing regulatory agency built in Gloucester for the tax revenue it would earn the city. Fishery regulation is a growth industry. Fishing is not.

Fishermen are increasingly raising issues of pollution and climate change. After all, if cod live within a certain temperature range and a certain salinity of water, and the ocean is becoming warmer and ice is melting into the sea, this should have a profound effect on cod. Most people in the fishing business have observed seasonal changes. Striped bass linger longer off Massachusetts past season, mid-Atlantic crabs shed their shells later, extending the soft-shell crab season, and the shad's annual charge up the Hudson River, once as reliable as a calendar, has been occurring unpredictably earlier. To date there have been very few studies on the impact of climate change on fish populations, and the number of studies on the impact of pollution have been minor compared to studies on the impact of fishing practices. Adding to the frustration of fishermen, when they raise these environmental issues, they tend to be dismissed by environmentalists, who imply that they are just trying to deflect blame for the crisis.

Bottom draggers from St. Peter's Square

Many Gloucester fishermen seriously believe that environmentalists are simply front men for big oil. The evidence they
offered for this was that many environmental groups, including
Oceana, which focuses on fishing issues, and the Cape Cod
Commercial Hook Fishermen's Association, got grants from
the Pew Charitable Trusts. This, however, was not surprising,
since Pew, which gives away hundreds of millions of dollars
yearly, was one of the leading philanthropic organizations in
America. Although it was founded in 1948 and financed with
shares of the Sun Oil Company owned by the children of the oil
company's founder, Joseph N. Pew, no one has shown a connection between the wide range of recipients and the interests of
big oil. The fishermen, in making this accusation, were referring back to their glory days when they took on big oil and
won. According to their scenario, the oil companies finance the

environmentalists to destroy the fishery so that they can get back Georges Bank for oil drilling.

Ever since the crisis on the Grand Banks, Gloucester fishermen have viewed the problem with an eye on the future. Sam Novello said, "There will be fish and the people who are left will do well." That is what most Gloucester fishermen believe. Most fishermen, based on centuries of anecdotal family experience, believe that fish are cyclical. They disappear for a time and then they come back. Rich Arnold said, "Fish run in seven-year cycles." The only disagreement among fishermen is the number of years of a cycle.

Government management believed that the fleet had too much capacity. So, while they would not ban bottom dragging, they did periodically offer to buy the vessels, giving fishermen a profitable way out. But Gus Foote urged fishermen not to take government buyouts. To him the government effort to reduce the fleet was an effort to diminish the power of local fishermen and eventually turn fishing over to a few corporations. "Everyone who sells is an opponent of LNG taken away," he said.

Fishermen remember this happening in agriculture—farms failing, farmers getting off of the land with federal assistance, and agro-industry replacing them. This was the nightmare of many Gloucester fishermen. In the early 1980s Gloucester was a dragger port with almost 150 bottom draggers. The old wooden side trawlers were being replaced with larger, more efficient steel-hulled stern trawlers. Then the calamities struck. In 1985 the World Court in the Hague settled a dispute dating back to the Revolution, declaring Browns

Bank, the northern edge of Georges Bank and an important piece of the Gulf of Maine, part of Canada's exclusive zone. Then U.S. regulators started closing parts of Georges Bank. Gloucester's offshore fishermen were being forced inshore. So in 1995, when the government offered to buy draggers from New England fishermen, a great many of the Gloucester boats were sold and destroyed, and the fishermen, to the great chagrin of the local artists, replaced them with smaller, less picturesque boats.

Gloucester had changed from an offshore dragger fleet to an inshore dragger fleet. But in 1998, when the government closed Jeffrey's Ledge, a 33-mile shallow strip that winds from Cape Ann to Cape Elizabeth in Maine, inshore dragging became limited. Jeffrey's had been a two- or three-day trip. Now few had large enough vessels to go back out to Georges Bank.

The fishermen who have survived have done so with a wily ability to adapt to a constantly changing set of rules. Many turned to gill netting, an old fishing technique where a net is anchored in the sea and fish swim into it and get caught in the holes in the mesh, literally grabbed in the gills by the net. Once light monofilament netting came on the market, gill netting became a lot easier, because lighter netting meant the boats could be smaller. Gill netting also had the advantage of being passive. The fisherman left the net at sea and came back for it later, and it continued fishing while he was onshore using up neither days-at-sea time nor increasingly expensive fuel. Draggers, on the other hand, have to run their engines and burn fuel to fish. A large dragger burns between eight hundred and one thousand gallons of fuel in a typical operating day.

But there is an environmental problem with gill nets. They're wasteful. The nets fish indiscriminately, and sometimes they break away from their moorings and roam the ocean, continuing to catch fish until the net is so fish-laden it sinks to the bottom to become food for predators.

Some fishermen keep changing not only gear but also target species. Once a particular stock becomes greatly diminished, the entire food chain becomes altered. The old adage—that nature abhors a vacuum—is particularly true in the ocean. When the cod population declined, other species moved in to take its place in the middle of the food chain. Some fish that were eaten by the cod now flourished, but then the species that they sought for food would become hard-pressed and that population, too, might collapse, from lack of food. The entire food chain changes, and the question for fishermen becomes, "What will be left that is marketable to catch?"

The Grand Banks moratorium, according to the Canadian government, was to last two or three years, to give the cod stocks a chance to recover. The fishermen were given assistance—"the package," to hold them over "until the cod came back." But cod stocks still have not recovered and the cod fishermen have become snow crab fishermen.

Gloucester fishermen have looked for new species and new markets, too. Mike Fogarty, a biologist at the Northeast Fisheries Science Center in Woods Hole, studied what he termed "a radical change in the ecological structure." In the past three decades, dogfish and skates, once called "trash fish," started to become increasingly commonplace. In 1973 cod, haddock, flounder, and other ground fish made up 70 percent of stocks on Georges Bank. Skates and dogfish were 22 percent. By the late 1980s, 15 percent of stocks were ground fish and 74 percent were dogfish and skates.

In the 1960s, large quantities of skates, which were at first inadvertently hauled up in dragger nets, were tossed into bushel baskets. At a dollar a basket, they were sold to lobstermen, who valued them as bait because skates have an extremely tough skin and would remain strung up to the nail in the crosspiece of the lobster pot even after many greedy claws had pulled at them. New Englanders are not adventurous eaters. Without the influence of Mediterranean immigrants, they probably would never have learned to eat the black mussels that cling in abundance to their rocks. Skates do not even look like fish; their spiny, tough exterior is clearly inedible and the white flesh inside is stringy and dissimilar to other fish. But by the 1980s the word had spread to New York and Boston that the French eat skate, and therefore it was not only edible, it was gourmet. When it began to appear on the menus of fashionable restaurants, Gloucester became a skate port. But the species could not withstand this pressure and soon only a few skates were to be found in Gloucester nets.

Americans never caught on to dogfish, even when it was called sand shark, but they sold better in the 1990s after they were renamed Cape shark. Bleeding, slimy, sandpaper-rough Cape shark started filling the open decks of gill netters. Much of this catch was shipped to Britain, where it ended up in London fish-and-chips replacing the traditional cod that was increasingly hard to find. The Gloucester fishing industry attempted somewhat unsuccessfully to promote the fish to local consumers. The eighth edition of *The Taste of Gloucester*, the 1976 surprise hit cookbook by the Fishermen's Wives of Gloucester, included a "bonus" section with four recipes for "Cape Shark, formerly known as Dogfish," including the following with a clear Sicilian influence:

CAPE SHARK SOUP DELIGHT

1 pound fish (Cape Shark)
½ cup diced celery
¼ cup stewed tomatoes
1 large potato cut into cubes
¼ cup olive oil
1 tablespoon parsley
½ cup diced carrots
4 cups water

In a saucepan put onions, oil, celery, tomatoes, carrots, and potato. Add 4 cups water, salt and pepper to taste. Cook for ten minutes. Then add two more cups water. When it comes to a boil add the fish (cut into bite-sized pieces first) and cook for 5 minutes more.

But soon the regulators closed the Cape shark fishery on the grounds that it was killing too many slow-growing females and endangering the stock. Since then, fishermen have been complaining of too many dogfish breaking their nets and eating their catch, and in 2006 they were given a new quota so that their seas would not be overrun with this predator.

But no sooner was dogfish stopped than Gloucester fishermen discovered hagfish. This snakelike sea creature has barely developed eyes because it lives in the dark nooks of the sea and is so primitive that some biologists argue that it is not a fish. In Gloucester hagfish are called slime eels because their natural defense is their ability to quickly cover their skin in a thick slime. When out of danger they tie themselves into a knot and squeeze out of it to scrape off the slime. They also have a sneezelike reflex to clear the slime mucus from their head, and have two pairs of tentacle-like protrusions around a slot instead of a jaw.

Gloucester fishermen always hated these unappealing creatures because they are scavengers and sometimes would get to a catch before it was hauled to the surface, entering the bodies of dead fish and eating them from the inside. But then it was discovered that Koreans would pay high prices for slime eels because they use the skin to make wallets, purses, shoes, and other items. The Koreans also eat the flesh. But the animal reproduces slowly and could be overfished far more easily than fecund fish such as cod. The Koreans had so overfished their own hagfish that they were forced to turn to imports.

CLEARLY SOMETHING is going wrong. While there is much overfishing in the world, it has not been practiced by Gloucester fishermen for more than a generation. Vito Giacalone said, "We are limiting rock hoppers, gear is more bottom friendly, we are thinking of limiting length of cables. We might restrict to no more than five hundred horsepower engines, but," he added, "you know, the Gloucester fleet is not a very deadly machine. Few have more than five hundred horsepower."

And it was true. Gloucester had become a port of small vessels, not the home of tremendously destructive factory trawlers. By 2006 only three of the seventy bottom draggers in Gloucester were more than ninety feet long. The Gloucester fleet followed all the regulations that were handed to them. And yet the fish stocks were not getting healthy. Peter Prybot, who wrote about fishing issues for the *Gloucester Daily Times*, said, "The fishermen have sacrificed and it just gets worse." That was the source of their growing frustration.

The overfishing that was on the record was due to regulatory mistakes. A new term was invented for these errors— retrospective pattern. Jacqueline Odell, who worked for the Gloucester-based Northeast Seafood Coalition, which attempted to organize New England fishermen, defined retrospective pattern as, "When the results come in you can see that the original data was wrong." In 2002 the regulators said Georges Bank yellowtail was back and encouraged fishermen to target it. But the retrospective pattern showed that the stock assessment was wrong and fishermen had been allowed to take too many. Hence, the fishermen were overfishing.

Odell said, "It's like saying we told you to catch a hundred fish but really we should have said seventy, so I'm afraid you are overfishing."

The question is: If the fishermen are following the regulators and the regulators are listening to the scientists, and yet
the fish stocks continue to be depleted, who is getting it
wrong? Priscilla Brooks is a lawyer for the Conservation Law
Foundation and one of the attorneys for the 2000 lawsuit
charging that the regulators were not fulfilling their mandate.
When asked this question her answer was, "It's not the fishermen, it's the regulators."

The regulators have relied on the science. But how good is
the science? It has certainly always lagged behind the crisis.
Scientists have generally learned about a species only once it
has started to vanish. In the 1930s, when the British first
began to recognize that fish stocks were in precipitous decline,
scientists could not even tell the age of a fish—critical information for assessing the health of a population. Only a group with
a broad range of ages is likely to flourish in the long term and
the presence of older fish, which are larger and deposit greater
numbers of eggs, is vital.

Can scientists accurately measure fish stocks today? Fish
stocks are measured by cpu's—catch per unit. A small area that
is presumed to be a certain portion of the whole, is fished. The
catch in that unit is multiplied, and it is presumed to be the
total. Small trawls with fine mesh are also passed every year at
the same place at the same time to look at the number of surviving young fish, which is known as recruitment. But scientists argue about the meaning of recruitment and whether it is
an accurate measure of the future of the stock.

Fishermen, who lack scientific method, have many genera-
tions of valuable experience. Scientists do not have this experi-

ence and distrust the fishermen's unscientific approach. But the scientists do not always prove to be right. In Newfoundland in the early 1990s, it was the inshore fishermen, and not the government scientists, who understood what was happening on the Grand Banks. There is also some question about the nature of cod stocks. Scientists have said that cod is a nonmigratory fish with a range of five hundred miles. After that, it is a different stock. Sometimes this is clearly true. The fish in different stocks can even look different. The slim, greenish Grand Banks cod looks substantially different from the fat, yellowish Georges Bank cod. But in the British North Sea ports of Hull and Grimsby, there are dragger captains who remember the days before the 200-mile limits when cod catches were failing in the North Sea and they became long-distance fleets. Some of the old-timers say they sometimes followed the same cod stock from Iceland clear across the Atlantic to Canadian waters. The scientists said this was impossible; but of late some scientists have been tagging cod and finding them later far from what was supposed to be their range.

Vito Giacalone said, "I still believe the science is pure and not influenced by a hidden agenda. The scientists do a lot of great work. The environmental groups' purpose is to get more funding, but the scientists who work for the government do a lot of great work. But unfortunately I still believe that they don't know shit. When they look at all the computer-generated models and science, in twenty-five years they will see that it was an exercise in voodoo."

Daniel Pauly, a scientist at the University of British Columbia and one of the leading experts on fish stock assessment, identified several problems. The first was what scientists are increasingly referring to as "the dreaded phone call." The dreaded phone call comes to a scientist from an elected official, notoriously, a European one. The politician explains to the scientist that it is understood that the stock assessment indicates that it is safe to take only 20,000 pounds of the species, but the "social situation," which in reality is a political situation, requires allowing 30,000 pounds.

The reason this happens more in Europe than the United States, though it happens in the United States as well, is that European fishery management is a huge barter system. There is bargaining between sectors—do this for cod and we will do that for potatoes—and also between countries: if we give the Spanish some hake quotas we have to give Denmark some herring.

Pauly also said that the amount of fish caught is greatly undercounted. Illegal catches are not counted because no one wants to admit they exist; sports fishermen's catches are not included, and by-catch is often not counted. The reason for not counting the by-catch is that fishery management is organized by species. Pauly said, "Shrimpers catch cod, but the working group on shrimp does not send data to the working group on cod." Cod is not their concern.

One of the greatest problems in fishery management is this species-by-species approach. If cod stocks are down and herring stocks are up, these are not unrelated problems because cod feed on herring. But there is no one in charge of managing the ecosystem, which is also why there is so little study of the impact of pollution and climate change. Large holes can develop in the ecosystem without anyone's noticing. Scientists suddenly realized that many sharks were vanishing from the Gulf of Mexico and that the barndoor skate, a large raylike predator, was vanishing from New England. But no one had been watching because no fishery targeted these noncommercial species—they were being killed in by-catch. Pauly is one of many scientists who has called for "ecosystem-based management."

But the idea is slow to catch on. Pauly said, "In New England we find the scientists most stubbornly sticking to the single-species approach. It is part of a long tradition of always being wrong."

Scientists also have the problem that it is in their nature to give the complex parameters of things, hypotheses, and probabilities. Regulators and journalists want to translate this into something concrete. Murray-Brown said, "What we are doing in fisheries management is a different kind of science. They are looking for a conclusion and it doesn't take much to get it wrong. The scientists are forced by regulators to provide an answer. To buffer myself I want to say that it is based on science. But scientists don't want to give these answers. They want to talk variabilities and assumptions.... If you ask a scientist if cod is overfished or not the answer is, it depends."

In the November 3, 2006, issue of *Science* magazine, an international team of fourteen distinguished scientists based on both the Atlantic and the Pacific Oceans published a study titled, "Impacts of Biodiversity Loss on Ocean Ecosystem Services," which is clearly not the sexiest of titles, but it was announced with a press release projecting that all commercial

fish stocks will collapse by the year 2048. It did what press re-
leases are supposed to do, and the article got wide coverage,
appearing not only in the *Gloucester Daily Times* but also in the
New York Times and many other publications. Actually, this important study—reported in four barely readable pages of careful science-speak, complete with charts, graphs, and thirty
footnotes—made no such claim. The study did show how the
declining biodiversity in the ecosystem contributed to the collapse of stocks and made recovery ever more difficult. In other
words, the worse it gets, the harder it will be to fix. However,
the scientists also found that it was not too late and recovery
was still possible. This simply confirms an established concept
of biology known as the insurance principle—biodiversity, a
wide variety of species, ensures the stability and productivity
of the ecosystem. It was a mixed report, full of qualifications
and lacking in certainties because scientists don't believe in
certainties—they believe in variables. These scientists would
not be surprised if the fish stocks survived 2048—but if that
happens, it will be said that the scientists were wrong.

MIRACULOUSLY, IN THE FIRST decade of the twenty-first
century, Gloucester remained a major port, second in New
England only to New Bedford. This was partly because of lobsters. Gloucester landed more Gulf of Maine lobsters than any
other port. The Gulf of Maine is part of Gloucester's inshore
grounds, but not all of these lobsters were landed by Gloucester fishermen. The political might of lobster fishermen in

Maine forced their state to bar the landing of lobsters accidentally picked up in dragger nets. This catch, landed in Gloucester, including some ancient monsters too big for a lobster pot—thirty-pound lobsters that are more than a century old—greatly contributed to the total value of landings in Gloucester, which is the standard measure of the commercial importance of a fishing port. The percentage of the catch that was ground fish, the traditional Gloucester catch, was a diminishing portion of the Gloucester landings.

Even so the fact that Gloucester still had some five hundred working fishermen, and their annual catch made Gloucester the nation's tenth largest port, was neither a fish tale nor a Gloucester story but an improbable and remarkable story of survival.

chapter Ten

THE SEA AND THE SEASIDE

. . .

and they stopped before that bad sculpture of a fisherman

—*"as if one were to talk to a man's house,*
Knowing not what gods or heroes are"—

—CHARLES OLSON, "MAXIMUM TO GLOUCESTER,"
SUNDAY, JULY 19, 1960

IT IS HARD TO TRUST ANY ASSERTIONS. WILL MOST SEAFOOD stocks in the world be seriously depleted by the year 2048? Have the number of large fish in the ocean decreased by 90 percent over the past fifty years, as Ransom A. Myers, a respected scientist, reported? Are 31 percent of the 274 commercially important fish stocks in the United States overfished, as a 2002 U.S. government report asserted? Are 60 percent of the fish species studied by the United Nations Food and Agriculture Organization either fully exploited or depleted, as an FAO report claims?

Nothing is certain in the ocean. Fish that were said to be plentiful have suddenly disappeared. Fish that were said to be extinct have been discovered alive, most dramatically in 1938 when a coelacanth, a fish thought to have died out with the dinosaurs, turned up on the deck of a South African trawler. Humorist Ogden Nash called it "Our only living fossil." According to the 1988 Washington, D.C., conference of biologists where the word *biodiversity* was invented, there are only about 20,000 known species of fish, including sharks, rays, and lampreys, which is not impressive, considering there are 50,000 known species of mollusks, and 751,000 known species of insects. New species are regularly discovered. Not all of the new finds are tiny. In 1976 a hitherto unknown species of shark was discovered when the fourteen-feet-long 1,600-pound giant attempted to ingest the stabilizing anchor on a United States navy vessel near Hawaii. But species are vanishing as rapidly as they are being discovered.

There is one certainty. Something huge—a massive shifting in the natural order of the planet—is occurring in the oceans, and with it will come tremendous biological and social changes. This shift, the disappearance of species, is also happening on land. Mammals and reptiles seem to be vanishing. A study by California's Stanford University published in 2004 in *The Proceedings of the National Academy of Sciences*, predicted that by the year 2100 up to 14 percent of all bird species may be extinct. Also in 2004, the World Conservation Union predicted that 12 percent of bird species, one-fourth of all mammals, one-third of amphibians, and 42 percent of all turtle and tortoises were facing the threat of extinction. Since the Endangered Species Act of 1973 was passed by the United States Congress, at least one hundred species have vanished from U.S. territory.

In 1950, more than 90 percent of the fish caught by commercial fishermen were taken from the Northern Hemisphere. Today Peru has one of the most productive fishing grounds in the world, and the European Union is sending its ships away from its own tired waters to fish African waters, underpaying impoverished African countries for the rights to their finite resources—an old story. Just as the steam trawler sent fishing from the southern part of the North Sea to the northern part, overfishing has now sent fishermen scouring the world's oceans in search of productive fishing grounds.

And what is happening to the old ports like Gloucester along both American coasts as well as Atlantic Europe, by the North Sea, and the Mediterranean? A graph could be drawn. A piece of paper can be marked from bottom to top with increments of money indicating profits. The paper could be marked from left to right with decades starting in 1780. Two lines are drawn; one, starting fairly high on the page, is for commercial fishing. The other, starting at the very bottom of the page but steadily rising, is tourism. The two lines form an X that crosses somewhere in the 1970s. While commercial fishing has been declining, tourism has become one of the fastest-growing industries. Coastlines that have been defined since ancient times by fishing, fishermen, and fishing culture—towns like Gloucester—are becoming rare. The small fishing villages of Georgia, the Carolinas, and the Chesapeake have been turning to tourism, sportfishing, and recreational boating. So have many of the fabled California fishing towns such as Monterey and Half Moon Bay, not to mention Fisherman's Wharf in San Francisco. In New England there is still Gloucester, New Bedford, and Point Judith, Rhode Island, but the maritime ports of Stonington, Newport, Provincetown, Chatham, Rockport, Essex, Newburyport, and many of the towns on the Maine coast are increasingly devoted to yachting, summer homes, beachgoers, and vacationers; they have little left of the fishing industry. The once important port of Mystic, Connecticut, like the cod fishing center of Lunenburg, Nova Scotia, has become a museum.

At first glance it would seem that tourism and fishing could coexist well. Tourists, like artists, love working fishing towns. But as Harold Bell said about Gloucester, "People come here because they love it. Then they want to change it." As tourism grows, it competes with fishing for almost everything. In 2007 the Siasconset Beach Preservation Fund in Nantucket outraged fishermen by proposing to scrape up 105 acres of fishing ground for sand to replenish a Nantucket beach.

In the conflict between the interests of tourism and fishing, waterfront space becomes a vital issue. Yacht owners pay prices fishermen can't afford for harborfront mooring and dock space. Tourists come to the town because they are attracted to a fishing port, then they start complaining about things they find unsightly or "smelly." Geoffrey Richon, a Gloucester developer, made this comparison: "It is like people who build a condo across the street from a bar and then complain that the bar is too noisy; there's something wrong with that. They knew the bar was noisy when they built the condo."

Newport, Rhode Island, a clean and luxurious yacht basin, is often cited by Gloucester people as what they don't want their city to become. But they could just as easily point to their neighboring Rockport, which shares their "island" and is now a town of shops and boutiques, bed and breakfasts, and art galleries. Cape Cod has become too expensive for fishermen to buy homes, and the fishing villages of Maine are disappearing one by one, as the summer residents buy up the property.

Even the reordering of the species seems to favor tourism. Because the predator species have been reduced on Stellwagen Bank, sand eels are proliferating. In the summers, only a few miles out of Gloucester, sometimes with the twin lights of Thacher Island still on the horizon, two of the largest animals ever to live on earth, humpback and finback whales, feed. They circle below, making a net of bubbles to trap the sand eels as purse seiners do with their nets, and suck the water through their strainers near the surface. The calves play, careful not to wander too far from their mothers, shooting almost straight into the air, their black bodies bigger than most bottom drag-gers. They do this with only two motions of their powerful tail and then they crash back onto the water. A shadow is seen and the whale suddenly breaks the surface, a tall, fishy-smelling spout of mist rising from its blowhole. On a good afternoon you can see a half dozen whales at once. No one in Gloucester can ever remember seeing so many whales before, and whale watching—a profitable trade for ex-fishermen—has become a thriving attraction.

One of the great changes in marine ecology stems from the fact that humans are loyal to their class, not in the Marxist but the Darwinist sense. As members of the class *Mammalia*, we tend to have more sympathy for mammals than other classes of animal such as the four classes that we collectively refer to as fish. So while people have hunted sharks to the point where they are threatened with extinction, they fiercely try to protect whales and seals.

Of course, this has not always been so, and these popular creatures were also driven to near extinction. But for more than a century, the concept of nature conservation has largely been about protecting mammals. Today it is far better to be a whale or a seal than a fish. The biggest problem for these marine mammals is that their food, fish, is getting scarce. Fishermen, as opposed to other types of *Homo sapiens*, hate seals. Seals eat huge quantities of fish, with a preference for the valuable ones, especially cod. And they eat wastefully, often taking only one big bite, usually from the belly where there are no bones.

Retired Gloucester fishermen Rich Arnold recalled, "When I was a kid, if we caught a seal we cut off its nose and took it to City Hall and got five dollars." But anyone caught cutting off seal noses today would have serious legal difficulties. And so the protected mammals are flourishing, especially harp seals and harbor seals. They even wander onto Gloucester beaches, where bathers, by law, are not allowed to approach them.

In the fish-processing operations along the Gloucester waterfront, there are glass tables with lights shining from underneath. These tables were specially designed to spot worms in cod fillets. These worms are a parasite that lives in seals. The more seals, the more worms. They are removed from cod with tweezers, and each worker has a pile of worms on the table. In recent years, these seal worms have also been spreading to haddock and flatfish.

A large seal colony has established itself where it has never been before—in Chatham Harbor on Cape Cod. Tourists are thrilled and boats are available for a closer look at the sleek-backed cavorting intruders. But for fishermen, it was hard to look at this teeming, hungry crowd and not connect it with the disappearance of inshore cod, which had forced them to voyage as far as one hundred miles offshore to find cod, a dangerous proposition for fishermen who work with light crews on small fiberglass boats. And weren't the seals also the reason for the increasing incidence of worms lowering the value of deluxe Chatham Cod? No one wanted to hear the fishermen's complaints against the seals. Cape Cod has a tourism-based economy now, and tourists liked the seals. The best hope for these fishermen was that the presence of all these tasty barking morsels will come to the attention of great white sharks, which might arrive to feast on their favorite food. The only fatal attack on a swimmer by a great white shark in New England history was in Cape Cod, in Buzzards Bay in 1936. The great whites are still out there, and if they start feeding on the seals, that would drive away both the tourists and the seals.

The one commercial fish species to have made a thorough comeback is striped bass. Striped bass are elegant, silvery, streamlined, hard-swimming fish with succulent flesh that is white and flaky like ground fish, but rich and flavorful like midwater species. They spawn in freshwater rivers in the South but also ones as far north as the Hudson and they swim out in the sea. While they don't travel to any New England rivers, they make their summer home the sea off New England, especially off Rhode Island and Massachusetts. There, to the joy of both commercial and sports fishermen, enormous striped bass are found. They were a plentiful fish until 1974, but then they started to vanish. By 1980 striped bass was gone as a commercial species. As a replacement, farmed striped bass went on the market. The farmed fish only resembled the wild one in that it had stripes. It was much smaller and had a pointed head and a different shape, and, as always happens with farmed fish just as with farmed mammals, had a completely different flesh than the fish that swam wild and foraged for food. But this was all that was left. Striped bass had vanished at least once before, in the nineteenth century. Now with overfishing and polluted rivers, it seemed that this was the end.

But with the help of the 1970 Clean Water Act, rivers were being cleaned up and states from Maine to Virginia severely restricted fishing. And the striped bass returned; by the early 1990s they were plentiful again. Dave Preble, a Rhode Island sports fisherman guide, wrote in his book *The Fishes of the Sea*, "Getting the stripers back, when no one really believed it would happen, was a rebirth, an affirmation of our own continuity."

But striped bass is as much a sportsman's fish as a commercial species and so it is part of the tourism world. Fishermen resent sports fishermen—the list of fishermen's resentments is long. They believe that sportfishing helps tourism to move in on their turf. They also feel that it is coddled by the government that regulates commercial fishermen. Sportsmen are often permitted to catch species and work fishing grounds that are off-limits to commercial fishermen. Commercial fishermen believe that government favors sportsmen because it wants wealthy vacationers to push them out. And that is why, according to fishermen, they managed the striped bass recovery so well. Now there is an overpopulation of stripers, a voracious eater of lobsters, and the lobstermen are complaining that striped bass are endangering their catch.

UNTIL RECENTLY, ONLY the hearty few vacationed in New-foundland, an island whose economy was based on cod fishing. The sparse population clung to the coastline in fishing villages built on pilings over the rocks and over the water. For a long time Newfoundland struggled with the loss of about a fourth of its young men in World War I. In France, on July 1, 1916, 685 Newfoundlanders were killed in a single attempt to charge out of a trench. This loss of fishermen weakened the will of the small British colony for independence, and in 1949 a bitterly contested referendum to join Canada instead narrowly passed. The argument may have finally been settled in 1994, when the moratorium put the island's fishermen out of work and they survived through assistance from the Canadian government. Even then, inshore fishermen pointed out that it was because of the Canadian government that the cod stocks were fished-out in the first place.

Since 1994, hard-pressed for an economic base, the rugged rock-bound island "in the middle of the Atlantic," it is often said—a slight exaggeration—has been increasingly turning to tourism. Ten years after the moratorium Newfoundland and Labrador, its more remote province-mate, were receiving about a half million visitors a year, which was about the size of New-foundland's population. They came to see where Cabot landed, look for traces of Vikings and Basques, visit two national parks, and witness a culture and way of life that may be disap-pearing. Suddenly landmark buildings were being discovered around every corner and the repair and maintenance of historic sights provided work for fishermen. Parks Canada also thought they would be helping fishermen when they proposed turning a one-time inshore fishing ground, Bonavista Bay, into an aquatic reserve for tourists. This is one of fishermen's most dreaded scenarios—that their boats will end up in museums, and they will end up guiding tour groups on their fishing grounds. Will fish suffer the same fate as the big game of Africa, preserved only in a few areas as a tourist attraction? The Bonavista Bay fishermen mounted such a vociferous opposition to this plan for their future that, when the bill for the park was in Parliament, the project was dropped.

The cod did not return to Newfoundland and life changed. Where there had been cod there was now crab. The fishermen were not certain if these crab had moved in because of the absence of the predator—cod—or if they were always there and no one had cared until the cod was gone. Inshore fishermen who had been getting eighteen and a half cents a pound for cod were now getting $1.60 Canadian for crab. Gone were the thirty-foot open-deck skiffs from which the inshore fishermen hand-lined or trapped cod—practices that required great skill and yielded modest catches, but which could have gone on indefinitely if huge bottom draggers had not swept away the stock offshore. Now the inshore fishermen dragged up their skiffs to lie in the weeds and bought bigger boats to go farther out and set baited traps, which they hauled back up on a winch. The offshore fishermen started crabbing, too. The draggers removed the huge spools of net from their sterns and rigged winches on one side for hauling crab traps. The fish processing

Abandoned cod skiff in Witless Bay, Newfoundland

plants became crab-processing plants. But it was a short sea-son—about two months in the summer and only 25,000 pounds of crab allowed for each license.

In 1994 these proud cod men had scoffed at the idea of fishing for crabs. But they decided it was better than giving up fishing. "It's boring," complained Bernard Chafe, an inshore cod fisherman from a family of cod fishermen in Petty Harbor, a little fishing village near St. John. "Hand-lining cod was a different challenge every day. This is the same every day. There's no challenge. But it's a living. Crab saved us," he said. "If it wasn't for crab, I'd be in Alberta." Many of their children have moved to Alberta and British Columbia, looking for a future. Bob Chafe, another Petty Harbor fisherman of no relation to Bernard, said that his son had stayed and was working on an oil rig. "He would love to be a fisherman," said Bob Chafe, "but there's no fish."

Yet a new cod industry had emerged in Newfoundland. The grocery store in Petty Harbor had the words "Souvenirs" and "Gift Shop" painted on it. Every little fishing village started selling a few souvenirs to visitors. In Petty Harbor and a lot of other villages there were only a few shelves in the back of the grocery store. What did they sell tourists? Cod: cod hats, cod T-shirts, cod-shaped chocolates, cod-shaped cookies, cod ornaments and sculptures, and cod business-card holders. One line of cod cookies was labeled "Endangered Species." The restaurants that catered to tourists rarely served crab. They brought in cod to sell because when tourists go to Newfoundland they want cod. In Bonavista, where tourists go because of the legend that John Cabot first made landfall on that spot, they even constructed fish flakes—platforms of rough-hewn timber like the stands once used for curing salt cod. But there were no cod to cure on these flakes. The only cod were the few that people are allowed to catch for home use that locals sold, not always legally, by the side of the highway.

THIS TENSION BETWEEN the tourism and fishing industries—really a struggle for the character and culture of coastlines—can be seen in much of seaside Europe as well. England has a long history of seaside tourism that probably, like overfishing, began on the North Sea. The earliest use of the word *seaside* in British English dates to 1797, at which time using the coast, the home of the fishing industry, for recreation was a novel idea. Even fishermen were not interested in the idea of a sea view and they built their homes as far inland as possible to be safe from the wind and the dangerous water. A 1787 guide to Scarborough, on the North Sea north of the Humber said, "the approach to sea ports are seldom particularly beautiful."

The first seaside tourism in Britain developed along the North Sea, especially at the mouth of the Thames, where famous oyster beds were struggling toward exhaustion, because it was the seaside closest to London. Margate, a small fishing village at the tip of Kent, not far from Normandy, was one of the first to receive visitors. Londoners started coming in 1765 on the return voyage of sailing ships known as hoys that cargoed wheat to London. By the early nineteenth century they were carrying twenty thousand Londoners to Margate in a good year. The voyage took a day, and in a stiff wind the hoys sometimes missed their port and blew the visitors out into the North Sea. Soon Margate offered "bathing machines"—small rooms by the shore in which a customer would take a quick cure in seawater.

Scarborough, a center of the herring trade since the Middle Ages, was another of England's first "seaside" resorts because, in addition to herring, it had mineral springs, and health spas with mineral springs were popular in the seventeenth century. Scarborough was unusual in that most such spas were inland. Being on the coast was not at first seen as an added attraction, even though a 1660 book by a Scarborough resident claimed that both bathing in and ingesting seawater had health benefits. This led a number of medical authorities to recommend bathing in the sea, including a Dr. Peter Shaw, whose own book included instructions on sea bathing in Scarborough. Originally sea bathing was done not for recreational but for medicinal purposes. People were sent to the seaside by doctors, and it was believed that the colder the seawater, the better. Winter was the peak season, and bathing was usually done early in the morning. By the second quarter of the eighteenth century, Scarborough, the former herring center, was becoming the bathing spa of England. The men were taken out a short distance in small boats called cobbles, from which they could swim naked. The women bathed from the beach in "gowns," which they changed into in small dressing rooms along the beach. By the early nineteenth century, fishing ports such as Yarmouth and Lowestoft were competing with Scarborough to draw bathers.

As in Cape Ann, what started as an exclusive experience for the aristocracy soon admitted the upper middle class and eventually even the unsightly masses. By the mid-nineteenth century, bathing at the seaside was no longer a cure, it was entertainment and relaxation—a vacation. And then seasides began to provide other entertainments. The seaside donkey ride that originated at Margate started to appear all along the English coastline. The idea of bringing children to the beach did not take hold in England until the Victorian age, and British resorts did not allow men and women to bathe together until into the twentieth century. "Mixed bathing" was something one had to go to the continent to enjoy.

At first these beach resorts did not encroach on fishing villages because fishermen preferred to build their villages away from the shore. But as the resorts became more popular, as the society became more mobile with trains and then cars, the resorts spread and overtook the villages. With increased mobility, English vacationers—not only Londoners but also people from the industrial Midlands—started traveling westward to the coasts of South Wales, Devon, and Cornwall.

JUST AS BEACH tourism first developed in England on the closest coast to London, and in Cape Cod and Cape Ann because of their proximity to Boston, in France seaside tourism began with the beaches and fishing ports closest to Paris, along the Normandy coastline. Soon the tourists were drawn to farther ports with more exotic cultures that the French state had tried to absorb—those of the Basques and the Bretons. Both were great seafaring people whose many centuries of taking cod from the waters off Newfoundland, *Terre Neuve*, was legendary. In the early nineteenth century, trips to the Breton seaside to see the colorfully dressed fishermen and salt workers became fashionable for Parisians, popularized by writers such as Balzac.

Later in the nineteenth century Parisians discovered the French Basque country, where almost every cove was a commercial fishing harbor. Travel writing was in vogue and Stendhal, Gustave Flaubert, Victor Hugo, and Prosper Mérimée all wrote of their trips to the Basque coast. Hugo enthusiastically compared Biarritz to his three favorite seaports in Normandy, writing that the Basque port particularly reminded him of Normandy's Tréport, "with its old church, its old stone cross and its old port, swarming with fishing boats." Today neither port is swarming with fishing boats. Biarritz completely lost its maritime identity and became one of the first large-scale beach resorts, along with San Sebastián on the Spanish side and a number of towns along the fishing coast of Normandy. Biarritz is a bit faded today and vacationers have moved on to other spots, but that has not made the fishing industry return. As Geoffrey Richon warned of the fishing industry in Glouces-ter, "Once you let it go, you can never get it back."

Along the entire Atlantic coast of Europe, including old traditional ports from where great sailing barks once landed Grand Banks salt cod, the declining fishing industry is under pressure to cede waterfront to the growing tourist industry. Some of the old fishing ports such as Ile de Re, San Sebastián, and Motriko, a popular resort for people from Bilbao, ceded most of their waterfront long ago. A few ports, such as Vigo in Galicia, a Celtic region in northwestern Spain, remain major fishing ports with the help of more than $100 million in subsi-dies from the Spanish government in addition to about $2.2 billion in fishing subsidies paid out annually by the European Union. Most fishing nations subsidize fisheries, recognizing the cultural and social importance of their survival. Japan, which pays out more than $2 billion a year in subsidies, is the leader in this practice, but the United States also pays out $1.3 billion to fisheries. All of these figures are subject to interpre-tation, there being wide debate on what constitutes a subsidy. Estimates of government subsidies to fisheries worldwide range from $20 billion to $50 billion. What percentage of this money goes to support destructive fishing practices is also debatable, but clearly billions of dollars are spent annually by the governments of the world to perpetuate modes of fishing that cannot be perpetuated indefinitely because they will destroy the fish stocks.

SOME PORTS STRUGGLE ON, like the rugged Basque port of Bermeo, where the town is built up on the mountain and tumbles down with paths and stairways to a working waterfront.

Inner harbor, St. Jean-de-Luz

Along this coast are some of Europe's oldest long-distance fishing ports, ports that may have sent fishermen to North America even before America was "discovered."

The French Basque town of St. Jean-de-Luz is one of the more successful attempts at blending tourism with commercial fishing. A famous old fishing port, the men once vanished for months every year to catch Newfoundland cod, but in the mid-nineteenth century they left the Grand Banks and focused on an inshore fishery in the Bay of Biscay, catching tuna, anchovies, and sardines. It still had an important tuna fleet that docked in the inner harbor of this expensive tourist town full of hotels, restaurants, and boutiques for summer visitors and retired Parisians. There were so many French people in town that locals, who had deep Basque roots, were worrying about the dilution of their culture. The Basque ports struggled with the rise of tourism and the decline in fishing. On the Spanish side there was a quasi-autonomous Basque government that concerned itself with the preservation of traditional Basque activities, but on the French side, the French government promoted tourism, calling the area *La Côte Basque* and trying to make it a playground for French people. Many Basques understood this as a direct threat to their traditions, and tourism-related buildings were vandalized at night when they were unoccupied. A message was written on their walls in the ancient Basque language, "Basqueland is not for sale."

Tourists who got excited about being in a working fishing port were offered a fishing trip. A boat took a dozen or more out of the harbor, past the protective seawalls into the choppy Bay of Biscay. The guests were each outfitted with a rod and spinning tackle heavy enough to land a twenty-five-pound bass. Instead, they caught tiny red rockfish between one and two inches long, barely enough to bend the rod. Catch-and-release did not appear to have caught on. The guides distributed small plastic bags in which the fishermen could store their tiny catch. They even offered recipes for the pathetic mini-fish, insisting that the head meats, which presumably would be extracted with tweezers, were particularly succulent.

Today the two leading French Basque ports are Bayonne and St. Jean-de-Luz. Known in Basque as Donibane Lohizune, St. Jean-de-Luz is a seventeenth-century port of rare beauty wedged between the Basque Sea and the blue silhouettes of the Basque Pyrenees along the Spanish border. It struggles to preserve its Basque culture, but the one thing that keeps it from being simply a resort for the French is the fishing fleet painted in the red, green, and white Basque colors and flying the Basque flag in the wonderfully sheltered inner harbor in the center of town. If the town loses that fleet it will lose its soul. The fleet is kept away from the tourists. The town is built at the mouth of the Nivelle River. One side, St. Jean-de-Luz, with its shops and restaurants, is for the tourists, the other, Ciboure, is home to the fleet, the early-morning fish auction, and most of the stores and offices related to the fisheries. Tourists are discouraged from visiting the fish auction.

The St. Jean-de-Luz/Ciboure fleet is a blend of modern stern trawlers and small, old-fashioned boats netting sardines. The average age of their boats is twenty years, which for Europe is an old fleet. The Basque fishing boat, marked by the graceful curve of its side rails dipping toward mid-ship, was outfitted for seining sardines in the winter and pole-fishing bluefin tuna, their most valuable catch, in the summer. This catch was supplemented by seining anchovies, which were so plentiful they were also used to bait the lines for bluefin tuna. Tuna, sardines, and anchovies are staples of the Basque diet in this area.

Traditionally, St. Jean-de-Luz fishermen used a rod and line to catch bluefin tuna, but by the end of the twentieth century, much of it was being caught by dragger nets. The result was that although the season lasted from June to October, the fishermen usually caught their European Union–allowed quota by August. The Spanish Basques caught tuna from the same stock in the same places throughout their season because they caught it more slowly by hook and line. This should have been a lesson for the St. Jean-de-Luz fishermen, who saw their cousins' catch sold in town once their quota was used up. Furthermore, in the fish auction in Ciboure, as in Gloucester, the line-caught fish fetched higher prices because it was in better condition. But the fishermen tended to miss the lesson and grumbled, like the British, that the European Union fishing policy favored Spain, which made as much sense as the Gloucester fishermen's dismissing the lessons of Chatham Cod because they thought big oil money was backing them.

Tragedy has already struck the Basque sea. The anchovy stock completely collapsed in 2007, and anchovy fishing—about a tenth of the St. Jean-de-Luz catch—was no longer permitted. This was like Gloucester without cod. In the picturesque central market of St. Jean-de-Luz, where superb local produce was sold at high prices to locals and to the affluent Parisians who visited or had moved there, a long-standing stall sold nothing but local cured anchovies. The stall was still there, but now all their anchovies came from Peru—or at least they will until that notorious maritime slaughterhouse collapses. In the meantime, like the cod of Newfoundland, St. Jean-de-Luz anchovies are only a memory, recalled in anchovy-shaped chocolates sold to tourists who cannot taste the real ones.

THE MEDITERRANEAN WAS one of the first bodies of water about which total destruction of fish stocks was discussed—by the French ecologist Jacques Cousteau back in the 1970s—and there today the process is even further advanced.

One of the famous traditional fishing towns in French Catalonia is Collioure. For many centuries there were two sources of income in Collioure—anchovies and wine. The wine, Banyuls, is dark, spicy, and sweet and goes perfectly with the anchovies, which are salted. Economically it was also a perfect relationship, the anchovy season starting in May when the vines were all tied up and had their shoots and could be left to grow. The men went to sea and the women mended the nets and sold the catch. Every family had its own vineyard and its own boat, called a *catalan*. A catalan was a wooden-hulled boat with a shallow enough draft to glide over the rocks of Collioure Harbor, powered by lateen rigging, a gracefully draped triangle of sail with a cross boom jutting up at sixty degrees. It was a design that dated back to the ancient Phoenicians, but the brilliant patterns that were painted on the boats in primary colors were Catalan.

They fished anchovies from May to October and then it was time to harvest the grapes. Since the Middle Ages, Collioure anchovies had been considered the best in the world. They are smaller, leaner, and more flavorful than the Basque anchovies of the Bay of Biscay. In the Middle Ages, Collioure had been famous for its sardines and tuna as well, but those fish stocks had vanished from the range of catalans centuries ago.

The bright catalans attracted one of the greatest colorist schools in the history of painting, the Fauvists. In May 1905, just as the families were turning from the vineyards to their colorful boats, Henri Matisse arrived. He invited his ten-years-younger friend, André Derain, to join him, and Derain arrived in July. There, they fashioned a vision that altered the course of twentieth-century art and attracted many other painters to the little fishing village. Matisse wrote later of his time in Collioure, "All I thought of was making my colors sing, without paying any heed to rules and regulations." Among the most important of the Collioure Fauvist paintings, compositions of pure bright colors, was Derain's 1905 painting of fishing boats and Matisse's painting of the boats from a window.

As anchovies became harder to find, fishermen needed larger nets to haul in the same catch, which required larger boats that would go farther out, but large vessels could not get in and out of Collioure's picturesque but shallow harbor. The fishing that attracted the painters is gone now, and the painters went with them. Banyuls is still produced, but the principal economic activity—the one that has replaced fishing from May to October—is tourism.

ON THE NORTHERN COAST of Sicily, west of Palermo, in the series of little fishing ports that gave Gloucester half its fishermen, there are fewer and fewer fishermen and fewer and fewer boats.

It's been a long, slow process for impoverished Sicily. Giovanni Verga, in his classic 1881 novel *I Malovoglia*, wrote about the brave fishermen of eastern Sicily going ever farther into the sea, powered by oars and sails, searching for an ever more meager catch because there were too many fishermen. The fishing has steadily grown worse.

On the other side of Sicily, in the town of Bonagia, near Trapani, fishing itself has been turned into a tourist attraction. The *tonnara*, the traditional Sicilian tuna hunt, began when the Phoenicians first understood that huge bluefin tuna pass by the western coast of Sicily at spawning time, May and June. Today bluefin tuna are one of the most hard-pressed species, owing in part to the extraordinarily high prices the Japanese will pay for a bluefin in good condition. It is not the *tonnare* that have destroyed the great schools of bluefin, but more modern techniques. Nevertheless this ancient fishing technique, far too difficult to be capable of depleting the tuna stocks, remains only in two places, Bonagia and nearby Favignana. The one in Bonagia is owned by a local tuna-packing company, which says it makes no money on it at all but keeps it alive for its cultural value. The company, Castiglioni, like Gorton's of Gloucester, no longer even buys fish from its area except the few bluefin picked up in the *tonnara*. They import yellowfin and pack it.

A *tonnara* is a colorful fishery. It involves about 120 fishermen led by one with the Arabic title of *raiz*. They go out to sea singing a song in Arabic to invoke the gods, "Cialomay." The *tonnara* begins in March, when the fishermen start working at the Bonagia waterfront. The *tonnara*'s processing plant and spotting tower for sighting the arrival of the bluefins is no longer in operation. It was converted into a hotel.

The hotel guests can listen to the singing, some in Arabic and some in Sicilian dialect—which can sound quite similar. Once at sea, the fishermen anchor a net 150 feet high and four-and-a-half miles long to the ocean floor. The tuna, trying to turn south to the straits between Sicily and Tunisia, instead hit the net and are forced to swim along it through a series of "rooms" until, exhausted, they are trapped in the final room and hauled to the surface, where fifty-five men are waiting to spear and gaff them. This final act, so bloody the water turns black and the whitecaps glow scarlet, takes place far enough out to sea so that the tourists in the hotel don't see it. These powerful fish can weigh as much as one thousand pounds and even exhausted it takes a considerable fight to land them. Of late, though, like most fish, they have been much smaller because as a fish population becomes stressed they start reproducing earlier and turn into a smaller species. The largest size classification for cod at the Boston market is now so rare in New England that it makes the newspapers when one is caught.

Cod landed on the trawler Kingfisher, *1920s*
(PHOTO BY GARDNER LAMSEN, COURTESY OF THE CAPE ANN
HISTORICAL ASSOCIATION, GLOUCESTER, MASSACHUSETTS)

The tourists in the hotel don't see the epilogue to the *ton-nara*, either—bluefins being packed and shipped to Japan, sold at the high prices the marketplace has deemed worthy of an endangered species.

IF YOU ASKED in Gloucester where to find her sister city in Europe, many people would probably say Terrasini, the original home of so many Gloucester families. But in truth there is no town on the Atlantic that more resembles Gloucester than the Cornish port of Newlyn in southwestern England. Like Gloucester, Newlyn is near the tip of a peninsula with outcroppings of granite, and has a sheltered harbor to the south in the lee of the wind. It is celebrated for its painters, artists, and metal workers, has now-defunct granite quarries, and is the last holdout on the peninsula—a rugged fishing town still, while its neighbors, Penzance and Mousehole, have turned to tourism. St. Ives, on the other side of the narrow peninsula, has also become concentrated on tourism. Land's End, the nearby landmark and western point of England, was purchased by an American company, which built a theme park. The first harbor on the coast, marked by a wall of granite, was called Pen Sans, which in the local Celtic language means "sacred headland." The sacred headland in Penzance can no longer be seen because a swimming pool has been built over it.

But Newlyn, like Gloucester, is a town that has been shaped by its many tragedies and is resistant to change. One of the most famous paintings of the Newlyn School, Walter Langley's

1884 watercolor, *Among the Missing*, shows the anguish and shock of a Cornish village receiving news of fishing boats lost at sea. The similarity to Gloucester has not been lost on the locals. When they decided to build a monument to the fishermen killed at sea, one of the first designs under consideration was a fisherman in oilskins at a wheel. But such wheels are found on Gloucester schooners, a type of vessel that was never used in Newlyn. The traditional Newlyn fishing vessel, a Cornish lugger, is steered by a tiller on the stern. In the end, in 2007, they settled on a fisherman casting a land line to come home.

The Cornish lugger was a small, two-masted, fore-and-aft rigged boat. Luggers were used throughout Great Britain, but the Cornish lugger, a particularly well-developed model, was distinct for its tapered bow-like stern, which allowed these vessels to pack tightly into the cramped Cornish harbors.

1900 postcard of luggers at the entrance to Newlyn Harbor

Cornwall, being Celtic and at a geographic extreme on the British island, felt a connection to Ireland and Wales, and even Brittany, but a distance from the rest of England. On the tip of the Cornish peninsula warmed by the Gulf Stream, Newlyn was the middle one of three fishing villages—Penzance, Newlyn, and Mousehole—lined up in a row in the lee of the wind along the eastern coast of Mount's Bay.

In the age of sail, Mousehole had an advantage because it was closer to the sea, and the fishing boats leaving had the wind more to their backs so that fishermen could get to sea far more rapidly from Mousehole than from the other two towns. From Mousehole they fished ground fish and shellfish, but their principal catch was pilchards, the local sardines, which much of the town was employed salting, packing, and shipping. After steam power came into common use, Mousehole lost its advantage, and many fishermen moved to the larger fleet at Newlyn.

Originally Newlyn was seen as a less desirable fishing port. Although well sheltered, the harbor tends to silt up and even today has to be dredged regularly. In the fourteenth century, Mousehole had ten times the fishing boats of Newlyn. But a century later, with demand for sardines growing, Mousehole Harbor was filled to capacity and Penzance had become more interested in commerce, and so a fleet started to develop in Newlyn—a fact which angered leaders in Penzance, who liked to think of their port as the mainstay for the region. By the early seventeenth century, the time when Gloucester was beginning, Newlyn had become a major fishing port, surpassing Penzance. Newlyn's harbor, unlike that of Mousehole, had room to expand and they steadily did so by building new piers. The most recent expansion was the Mary Williams Pier that stretches through the center of the harbor and was built in 1980.

Mousehole had one advantage left. Its perfect semicircular harbor enclosed by a granite wall, its waterfront jammed with the masts of fishing boats, was considered one of the most picturesque fishing ports in Great Britain, a great attraction for tourists. The poet Dylan Thomas, staying there in 1937, called Mousehole "the loveliest village in England." He didn't say "in Britain," because that would have included his native Wales.

Mousehole is still one of the most picturesque ports. The fishing boats are long gone but the tourists are still there. The town, like St. Jean-de-Luz, is filled with retirees and people with second homes, menacing the local Celtic culture. A local commented, "I don't suppose there are 125 real locals living here now." Among those real locals are only a few fishermen who work out of Newlyn. In 1931 the population of Mousehole was two thousand. By the year 2001, it was down to 607.

But they still have their traditions. The night before Christmas tourists and locals like to celebrate the memory of Tom Bowcock, a legendary fisherman, Mousehole's Howard Blackburn, who saved the town from starvation by braving the storms of December 23 and bringing back a load of fish. Unlike Howard Blackburn, no one actually remembers Tom Bowcock, or Beau Coc, as he was known in an earlier version, and the figure may trace back to Celtic legend. But the locals make "starry gazzy pie" to celebrate the occasion. The original dish contained pilchards stuffed with herbs, arranged in a wheel so that the tails stuck through the crust in the center and the heads burst through, staring upward in a circle around the edge of the pie. Today the Mousehole festival is centered around The Ship Inn. This is the recipe of their chef, Richard Stevenson:

PASTRY

> *300 grams plain flour*
>
> *150 grams butter*
>
> *Cold water*
>
> *milk or melted butter, for pastry glaze*

FILLING

> *400 grams undyed smoked haddock fillets*
>
> *200 grams pollack fillets*
>
> *200 grams whiting fillets*
>
> *200 grams ling fillets*
>
> *6 herring or pilchards/sardines, for presentation*
>
> *3 hard-boiled eggs, chopped*
>
> *Sea salt and freshly ground black pepper*
>
> *1 litre milk (or half milk and half water)*
>
> *1 small onion, peeled*
>
> *3 cloves*
>
> *Small bunch parsley—separate the stalks*
> *   and leaves, finely chop the leaves*
>
> *50 grams butter*
>
> *45 grams plain flour*

Make the pastry by rubbing the butter into the flour, and then adding enough water to bind the ingredients together. Chill in the fridge until required.

Cut the heads and tails off the herring (or pilchards/ sardines), and put aside. The cuts must be angled so that on the finished pie, the fish look upwards. The remaining herring can be filleted, skinned, and added to the fish mixture if desired.

Poach fish fillets in milk (or milk and water) with onion, clove, and parsley stalks, until just yielding to a knife. Drain off poaching liquor and reserve, and flake the fish.

Make a béchamel sauce, using the butter, flour and poaching liquid. Season, and add the flaked fish and the chopped parsley. Put this mixture into a large pie dish, and add the hard-boiled eggs.

Roll out the pastry, and use it to cover the pie. Arrange the herring heads and tails to look as if they are swimming through the pie. Cover the eyes with pieces of pastry during cooking, to keep them bright.

Glaze the pastry with milk or melted butter, and cook in a hot oven at 200°C (400°F) for 10 minutes, then at 180°C (350°F) for about 50 minutes, depending on the size of the pie.

In reality, inshore fishing is always bad in December and January in these Cornish waters because the fish move south to spawn. This was the origin of preserving the pilchards for winter, and some believe the entire legend of Tom Bowcock is a cautionary tale to remind Mousehole residents to put up fish for winter.

The export of cured pilchards has been an important business in this area since the early sixteenth century. By the eighteenth century, enormous quantities of French sea salt were imported to Newlyn and Mousehole for the production of cured pilchards for export. In past centuries, when the Catholic Church kept much of Europe eating fish, the courtyards of Newlyn, which today house little tea shops and workshops for crafts, were literally flowing with the bloody brine of fresh pilchards being cured.

Today there is only one producer left in all of Great Britain, British Cured Pilchards, Ltd. of Newlyn, which is also open to tourists as a museum. Margaret Perry, a local historian who recently moved into a new apartment in a former pilchard plant in Newlyn, offered this traditional recipe:

SCROWLERS

Descale and clean the pilchards and split open. Season well. Grease the hot plate or griddle, cook the fish quickly, one side and then the other. Years ago, these were often cooked over an open fire, indoors. The smell was unbearable. Excellent barbecue food, though!

MINING HAD ALSO BEEN a traditional economic activity in Cornwall. Newlyn had been a tin and copper mining town. Tin mining dated back to ancient times, and the town's skies were charcoal-smudged with the smoke from smelting until the operation closed in 1896. In the mid-nineteenth century, mining and associated industrial activities went into recession at about the time affluent people decided that they liked to be by the sea. Since fishing ports were considered dangerous and ugly, Penzance built a second town away from the port for tourists. This made Penzance, as opposed to Newlyn, an appropriate place for "the better people" to visit. The port prospered especially after 1859, when a direct rail connection was completed from Penzance Harbor to London. In 1844 a paved promenade along the sea was built in the center of town. And in 1866 the Queen's Hotel was built, providing sea views for its guests and obstructing the sea view for everyone else. Tourism grew and today it is the principal economic activity in Penzance.

But in Newlyn the railroad was seen as a different opportunity to compensate for the decline in mining. The railroad line that brought tourists to Penzance also shipped fresh fish from Newlyn. This made fresh fish, especially shellfish, an important industry with a large market beyond the local area. Newlyn was celebrated for its brown crabs, a seafood that travels particularly well.

Crab soup is a specialty of Newlyn. This recipe comes from Marie Harvey, the mother of the leading Newlyn shellfish company, a three-generation family firm, W. Harvey and Sons. Cream of crab is the meat from the crab's body, which is thought to be unpresentable because of its brownish color, but it is particularly rich in flavor.

CRAB SOUP

SERVES FOUR

2 ounces butter

1 tablespoon olive oil

Lemon juice

1 teaspoon curry powder

4 ounces cream of crab

1 medium onion

Approximately half pint milk

2 ounces white crab meat

1 level dessertspoon plain flour

Salt and pepper to taste

Chopped parsley

Melt butter with olive oil in saucepan. Add chopped onion and cook gently until tender, but do not brown. Blend in flour with the curry powder and mix with the onion. Gradually stir in sufficient warm milk to make the thickness of soup required. Add a dash of lemon juice. Mix in the cream of the crab, and lastly the white crab meat. Extra milk can be used if necessary. When serving, garnish with chopped parsley.

AS IN GLOUCESTER, the sights of a working fishing town in Newlyn drew artists. An art critic at a Royal Academy exhibition in 1888 wrote, "The Newlyners are the most significant body of painters now in England." They became even more prominent after 1899, when Stanhope Forbes and his wife, Elizabeth, founded the Newlyn School of Artists. Many in the group had studied in France and were struck by the similarity of Cornwall to that other Celtic peninsula, Brittany. Influenced by French Realists, Newlyn artists such as Walter Langley, Frank Bramley, Thomas Gotch, and Laura and Harold Knight, painted water, boats, and fishermen from the streets of Newlyn using flat-ended brushes for pronounced brushstrokes and thick paint in subtle colors. The Newlyn painters changed British art by rejecting the classical and romantic themes of Victorian and Pre-Raphaelite painters, and depicting instead "the common man," who in the case of Newlyn was usually a fisherman. Stanhope Forbes was popular with fishermen because he was known to pay them well for posing. To a fisherman, posing for a painting is an incredibly easy way to make money.

In 1888, John MacKenzie, a tall, handsome man in his late twenties, already known in the London art world, came to Newlyn, which had a tradition of copper making, and he started what he called the Newlyn Industrial Class. He taught young men, many of them fishermen, to make copper in the Arts and Crafts style, which was becoming popular—a response to what had been seen as the aesthetic mediocrity of industrialization. Newlyn copper, hand-beaten into designs from the back, a technique known as repoussé, often with fish or fishing themes, became famous and provided work for fishermen, a group with notable manual dexterity, in lean fishing seasons.

Newlyn even had its own version of the Folly Cove Designers, the Crysede Silkworks, which began in 1919 and was famous for innovative silk printing.

NEWLYN HAS DONE a better job than Gloucester of preserving its historic architecture, the old granite buildings of the fishing industry along the Combe, a street running up from the harbor, and ancient narrow lanes with granite houses whose ground floors were dedicated to curing fish. But, like Gloucester, Newlyn has kept a rough, blue-collar exterior. The feel is of a working port, not something finished and picturesquely preserved like Mousehole, or for that matter, Rockport. Newlyn did not even have electricity until the late 1930s.

After World War I, the British Government passed the Slum Clearance Acts, which were enacted to tear down working-class housing and build sterile new buildings with "modern conveniences"—homes, it was said, "fit for heroes." But they did not ask the heroes if they wanted new homes.

The people of Newlyn did not want to see their homes razed. Fishermen did not want to live far from the water or in a building with no place to store gear, and the fishermen joined forces with the artists in protest. Artists and fishermen are sometimes an odd coalition, but both groups have strong feelings about preserving the traditional. The protest culminated in 1937, when a 1919 Newlyn sardiner, the *Rosebud*, carried citizens with a petition to London. They had to sail to the North Sea on the other side of England in order to enter the Thames.

As they went upriver, horns were blasted, cranes were lowered, and crowds cheered from the bridges and the banks. The voyage of the *Rosebud* was featured in newsreels, and by the time the Cornish protestors reached the Ministry of Housing and Parliament, their petition was too famous to be ignored. Soon the rebuilding program was entirely dropped for the demands of World War II. Eventually much of the Newlyn "slum housing" became historic "quaint little cottages," much prized in the real estate market.

Like any fishing port, Newlyn has known many tragedies. In 1880 the *Jane* went down with all hands in full view of the dockworkers. Newlyn also had its own Howard Blackburn–style exploits, such as the seven Newlyn fishermen who in 1850 sailed the *Mystery*, a 33-foot mackerel boat, 12,000 miles in 116 days to Melbourne, Australia, with only one stop in Cape Town.

WITH FORTY ACRES of harborfront, Newlyn has become, along with Brixham in neighboring Devon, the biggest fishing port in England. The great North Sea ports have all faded. Distinct among all the Cornish ports, Newlyn, like Gloucester, has stayed stubbornly focused on fishing, has a working-class ethos, and wants a tourist industry only as long as it does not encroach on the fishery's space.

While Newlyn does have stern trawlers and bottom draggers for ground fishing, as in New England, ground fishing has been increasingly troubled and restricted by regulations.

Although Cornish barks crossed the Atlantic in the seventeenth and eighteenth centuries to fish for cod off of Newfoundland, Newlyn, unlike Gloucester, was not a ground-fishing port to begin with, but fished for mid-water species such as pilchards, herring, and mackerel with purse seines. In the nineteenth century, the Cornish started using drift nets, like unanchored gill nets. This greatly extended their fishing season from the fall run of pilchards, to winter herring, to mackerel from February to July. Newlyn fishermen had traditionally caught ground fish by hook and line, although sail-powered beam trawlers were used there in the nineteenth century. But, in the late nineteenth century, when the fishermen of Brixham were among the first to take steam-powered draggers to the North Sea, Newlyn fishermen refused to join in, saying that these vessels were too destructive. Newlyn got its first steam-powerered mid-water drift netter in 1908, when there were already more than two hundred such vessels in the North Sea. Part of the reluctance of Newlyn to embrace engine power seems to have been the cost of importing Welsh coal. Once the internal combustion engine became available in the early twentieth century, Newlyn fishermen started using engines.

Between 1930 and 1950, bottom draggers became an increasingly important part of the Newlyn fleet, catching ground fish, especially cod. Most of the bottom-netting vessels that remain today—expensive stern draggers and beam trawlers costing up to $1 million each—are owned by a single Newlyn company, the 200-year-old W. Stevenson & Sons, which owns thirty-five Newlyn vessels and claims to be the largest private fishing company in Europe. But the number of these large vessels is steadily declining.

Newlyn Harbor in 1950

One of the reasons that Britain, in the 1960s, was reluctant to join the European Common Market was that British waters

contained some of the best fishing grounds in Europe. In the 1980s, when Spain and Portugal entered the Common Market, both countries had great fishing traditions but little to offer in fishing grounds. They had lost "their" grounds—New England and the Grand Banks—with the establishment of the 200-mile limit. The British government had agreed to limit British fishermen to only 18 percent of fishing quotas in British waters. This was a bitter blow for the Cornish, who had a long tradition of protesting against outsiders fishing their grounds. Outsiders had included vessels from British North Sea ports as well as from neighboring Devon. Complaints had ranged from destructive fishing practices to a notable 1896 riot over fishing on the Sabbath. Now Newlyn was again in crisis. The arrival of Spanish boats was called an "invasion," and it was frequently pointed out that the Spanish had invaded Newlyn before, albeit in 1595. Europeans have long memories.

It is doubtful that the Spanish vessels, limited in number and in the size of their catches, greatly damaged Newlyn. There were many other newcomers, including the big draggers from Hull and Grimsby, that could no longer fish in Iceland and a large mid-water fleet from Scotland.

The British government, recognizing that with only an 18 percent share, its fishing needed to be reduced, launched a policy of "decommissioning," which is known in Gloucester as a "government buyback." Fishermen were paid to have their vessels destroyed. Since the program targeted large vessels it reduced far more North Sea and Scottish fleets, but in the heart of Newlyn's waterfront, too, on any given day a vessel could be seen being cut down to scrap. As in Gloucester, some left the fishing industry and others adapted, replacing large bottom draggers with small gill netters, and new species once thrown over the side of draggers, such as monkfish, were targeted. Newlyn survived with some five hundred working fishermen, when only about fourteen thousand remained in all of England. The 168 vessels of the Newlyn fleet included only five large vessels but more than fifty small one-man boats and fourteen old-fashioned beam trawlers. As in Gloucester, the switch to small vessels increased the fishing pressure on nearby inshore fishing grounds. Most fishermen worked from one-half mile to three miles offshore. Some fishermen caught mackerel, pollack, and bass with hook and line and landed them fresh and carefully handled. As in Gloucester and St. Jean-de-Luz, in the display auction every morning in Newlyn, these line-caught fish fetched the highest prices.

The great advantage of Cornish fisheries is that they catch

an unusual variety of species. A sixteenth-century survey of Cornish fishing reported more than forty-five types of commercial fish commonly caught, and this list is just as long today because five ocean systems, each with different temperatures and salinities—the North Atlantic, the Gulf Stream, the Bristol Channel, the English Channel, and the Irish Sea—all converge near the Cornish peninsula. Even a small vessel in a day trip will often bring in more than twenty species.

So when some stocks become threatened, the Cornish can turn to other fish. In 2007 the traditional mid-water catch, including pilchards, was down and the biggest catch in Newlyn was megrins, a blond European flatfish of mediocre quality, followed by monkfish and then pollack, none of which are traditionally important.

Newlyn fishermen, like Gloucester fishermen, have survived by operating small boats and keeping their options open. David Pascoe was from a Newlyn fishing family and from October to February he rigged his small boat for purse-seining pilchard. From February to March he went after ground fish, mostly pollack and ling. He would go after cod, but he was allowed to land only fifty kilos a month. After that, all the cod he caught were thrown over the side. His brother's boat was allowed to catch two hundred kilos of cod a month, which was the limit for vessels over ten meters. David's boat, the *Little Pearl,* was only three centimeters shorter than his brother's, but that put it under ten meters. He could put a strip across the stern to make his boat over ten meters. But then he would need a license for an over-ten-meter vessel and that could cost as much $160,000, if he could find someone willing to sell one. Since

1993, when every British fishing boat got licensed, with no new ones to be issued, licenses had become more valuable than vessels. This was true in the United States as well, which is why when Gloucester fishermen negotiated for a payment from the state of Massachusetts as compensation for losing fishing grounds to the natural gas project, the fishermen wanted to use the income to buy licenses and make sure they remained in the hands of Gloucester fishermen and not large, out-of-town fishing companies.

Newlyn fishermen knew their seas and complained that they were regulated by European Union officials in Brussels who didn't. Robin Turner, a tall, burly man who looked like the ex-rugby player he was, ran the fish auction early every morning. His family had been in the fish business since 1765, when, according to family lore, a Jewish butcher from Palestine married a Christian from a butcher family in Italy, and so many family arguments about butchering ensued that it was decided to change the family business to fish. Turner's father talked about a golden age of fishing after World War II, when the British government was promoting fish as a way of getting more protein into British diets, and Newlyn built huge trawling and drift-netting fleets, which caught enormous quantities of fish, with government grants. But that was the beginning of the trouble.

Today, Cornish fishing is largely regulated by quotas on species and by closing off areas. Days-at-sea restrictions have not been used very much because in those waters small vessels are limited in days by bad weather, which is what Turner calls "natural legislation." Turner said, "Natural legislation is more important to us than paper legislation. Paper legislation is done by people who don't understand how to fish." The Newlyn fishermen long advocated and finally got a 240-square-mile area closed to cod fishing in April and May during spawning season. But many disputes remained with Brussels about the management of the Cornish fishery. The fishermen do not understand why fish farming is encouraged when the farmed fish are fed wild fish. "Where is the logic," argued Turner, "in using four to seven tons of wild fish to make fish meal to produce one ton of farmed fish?" Even in medieval times, fishermen in Britain had protested the making of fish meal for animal feed as wasteful.

"The biggest problem," said Turner, "is convincing government of the value of regional management." Here, as in New England, fishermen hated the wastefulness of quotas that forced them to dump catch overboard. A 1989 law that permitted landing only 5 percent of by-catch meant that most of the nontargeted species were thrown overboard—a painful waste in an area where the variety of species makes it impossible not to have a large by-catch.

In 2007 the British government, after what they said was a five-year investigation, charged seventeen Newlyn fishermen and shipowners, including the biggest owner, Elizabeth Stevenson of W. Stevenson & Sons, with illegally landing fish that were over quota. Unwilling to throw away valuable fish that were already dead but were over quota, six Newlyn vessels had been landing more than their quota of cod, hake, and monkfish by labeling them ling, turbot, and bass—fish for which there were no quotas. The fact that it took five years to catch them may indicate how little regulators know about fish.

The fishermen did not deny having gone over their quota, but they argued that while barely eking out a living, they could not bring themselves to throw out valuable fish that were dead anyway. Drew Davies, one of what became known as the Newlyn 17, said that during one trip he had been forced to dump a thousand dead cod overboard. "There is nothing worse for a fisherman than doing that," he said. Another of the accused, Steve Hicks, a former policeman, told the London *Guardian,* "We knew we were doing wrong. But it wasn't done with greed. It was done to make a living." After the seventeen were convicted in the summer of 2007, a British government spokesmen had called the case "a major success in the control of overfishing," which was probably true from an administrative point of view. From a biological one, however, this seemed less certain. Another new form of overfishing had been documented: selling fish instead of throwing them away.

THE LOCAL FISHERMEN wanted government to pay fishermen to work with scientists, arguing that they would do it well and the supplementary income would help them to fish less. "But government doesn't trust fishermen," said Turner. "And I understand because many have abused the system. But we are not the dunces government thinks we are. We don't want to be the last bastard catching the last fish and feeling proud."

Crabber in the Cornish harbor of Cadgewith

As in Gloucester, the ground-fish catch had considerably diminished, but fishermen had made up for it with other catches. Ironically one of the biggest markets for Newlyn's catch was Spain. Spider crabs, which the British wouldn't eat and which, until recently, were regarded by Cornish fishermen as a pest, had become a new and successful catch because the French and Spanish would pay high prices for them. Even the British were embracing new species. A "pest" fish suddenly became trendy. David Pascoe said, "Someone famous decides he likes it and away it goes." Pilchards were back in fashion, too, since regional cuisine had become fashionable and restaurants had started calling them "Cornish sardines."

But shifts in species are only a temporary solution. We

can sadly learn to live without our favorite white-fleshed fish species, but if ground-fish stocks do not recover, if they have a drastically reduced presence in the food chain, the impact on the entire marine ecology will be huge and probably devastating.

Buyers from all over Europe go to the Newlyn fish market every morning. The many varieties and sizes of catch are sorted into red plastic boxes on the floor, and the buyers, pen and notebook in hand, follow Turner around the room while he auctions off box after box. Watching eight buyers clutching their pens and furiously bidding on a box containing one lone but handsome four-pound haddock—valuable because it was the only one brought in that morning, it was hard not to wonder if this was the future of commercial fishing.

chapter Eleven

SURVIVING ON THE MAINLAND

. . .

Gloucester too

is out of her mind and
is now indistinguishable from
the USA.

—CHARLES OLSON,
"DECEMBER 18," *THE MAXIMUS POEMS,*
VOLUME THREE, 1975

FOR THE PEOPLE OF GLOUCESTER, CRISIS IN FISHING IS A
familiar idea. The trauma for which their history had not pre-
pared them was the loss of their island. It started to happen
after World War II, at the very first stages of planning for a
federal highway system, including Interstate 95, which curves
around Boston to its west on its way from Miami to Maine. The
Boston stretch of I–95 was also to be the beginning of Route
128, a highway that was to link the South Shore with the North
Shore without going through Boston. The end of Route 128 is
East Main Street in Gloucester. Gloucester had to put up a traffic light for it, one of two in the city. Most intersections in
Gloucester are governed by such arcane rules that only the locals, and not all of them, know what to do. In 1950, long before
Route 128 was completed, a bridge was built over the An-
nisquam as part of the highway project. This was nothing like

the Cut bridge, which rises and falls to connect and disconnect Gloucester. This was a permanent bridge of steel and granite, one hundred feet in the air, high enough to allow tall-masted vessels to pass under it and wide enough for four lanes of traffic.

The bridge was named after A. Piatt Andrew, a World War I hero who was also the only Gloucester resident ever to serve in the U.S. Congress, although he was actually a native of La-Porte, Indiana, who moved to Gloucester at the age of twenty-seven, when he took a position as an assistant professor of economics at Harvard.

When the bridge first opened, Gloucester people enjoyed strolling across it for a pleasant walk over the wide Annisquam to and from the mainland. But gradually they realized that with this new bridge, Gloucester would no longer be an island. Andrew's way of life, living in Gloucester and working at Harvard, had now become much easier, thanks to the bridge that bore his name.

Boston had never been very far from Gloucester. On a clear day, high-rise buildings in downtown Boston reflecting the sun can

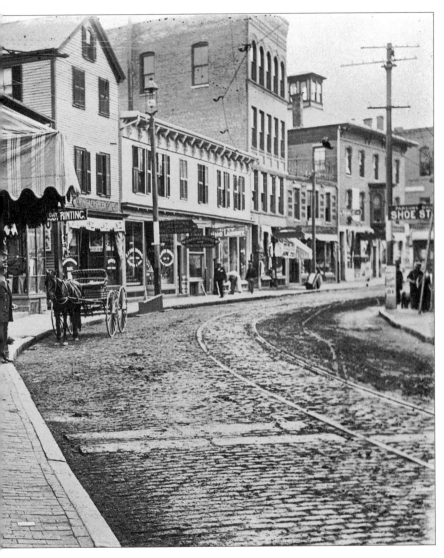

Main Street, Gloucester, in 1895. It looks similar today.
(COURTESY OF BODIN HISTORIC PHOTO)

be seen as a golden mirage across Massachusetts Bay from Gloucester Harbor. But now, Boston became a half-hour commute. This meant not only that it was easy to live in Gloucester and work in Boston but also that Bostonians could easily drive up for an afternoon on the beach. The idea of Gloucester becoming a Boston suburb was disturbing to most Gloucester people. The island was gone. No wonder the police had to rig antisuicide fences along both sides of the bridge. It had became the jumping spot of choice, not a sure leap but one full of symbolism and drama.

Some still can't accept the change. In 2005, Angela Sanfilippo continued to refer to Gloucester as an island. "That's why Sicilians are comfortable here," she said. The fishermen especially cling to this idea of the island. Tom Biacaleone said, "When I leave here there is no water around and I feel lost."

Ronald Gilson, born in the 1930s, as a teenager ran the water boat that delivered fresh water to the fishing boats around Gloucester Harbor. In 2007 he self-published a memoir, calling it *An Island No More: The Gloucester I Knew.* He wrote about how Gloucester used to be "a self-contained city, an island, literally: the ocean separated us from the outside world. We were a complete entity, supported mainly by our anchor industry—fishing. As a young boy, I thought this fantasy would go on forever, it was a magical time."

One of the first outward signs that Gloucester was no longer an island, something deeply felt by many locals but little understood by outsiders, was that the city's newspaper changed. Gloucester has a long tradition of having its own newspaper. The current paper, the *Gloucester Daily Times,* started publishing in

1888. It was a Gloucester-owned paper that covered Gloucester issues with local writers, including Charles Olson, regularly contributing. But in recent years, like most newspapers, it is no longer locally owned. By 2007 it boasted a circulation of just a few less than 29,000 readers, impressive for a town with a population of only 29,000. But it did this by covering all of Cape Ann, which seems reasonable for such a small area, but which meant that the *Gloucester Daily Times* was no longer the island's newspaper. It ran right-wing editorials that were not in character with local thinking and in its few pages covered the other Cape Ann towns.

Another sign that Gloucester was no longer an island was the real estate market. Tom Biacaleone reminisced about when he first moved to Gloucester from Sicily in 1956 and the waterfront along East Gloucester "was one long roof for Gorton Pew fish flakes and there was a mountain of Sicilian salt." Today that waterfront is valuable for other reasons. The days when the rich lived on the high grounds and the waterfront was for the fish and the poor are over. The harborfront in East Gloucester where the fish flakes once stood is prime property. So are the once-delapidated waterfront neighborhoods of Rocky Neck where poor artists used to live.

The painter Theresa Bernstein told the story of how she and her artist husband, William Meyerowitz, got their house in East Gloucester. One day they were out working together and Meyerowitz saw a scene he wanted to depict with watercolors, so he knocked on a door and asked for some water to paint with. According to Bernstein the woman in the house said, "Why don't you buy my house?" The two painters had no money other than Meyerowitz's earnings from the sale of several etchings to a wealthy man from Kansas. But the owner took them to a bank where they signed some papers. They started to walk out and the banker called out, "Mr. Meyerowitz, what about a down payment?"

"I don't have any money," Meyerowitz protested. He had $40 left from the sale of the etchings and the bank took that as a deposit, and so Meyerowitz and Bernstein became people of property in Gloucester. But that was the Gloucester of the Great Depression, in the 1930s.

There was a time when Gloucester real estate seemed up for grabs. Since most of the residents were immigrant fishermen, it was an affordable town with affordable property. But it had the setting—the sea views and the light and, not least of all, the sight of working fishing boats coming in and out of the harbor.

Once the highway was built, and Gloucester became just outside of Boston, it was in danger of becoming, like Cape Cod, Penzance, or St. Jean-de-Luz, a fishing town where fishermen could no longer afford to live. One by one, fishermen, realizing the value of their humble homes, started cashing in, selling their houses and moving away. Given the way fishing was going, the real estate market seemed to some the best opportunity they were going to get. Some fishermen even stayed on and became real estate brokers.

EVEN THOUGH THE population of Gloucester had barely changed since 1950, when it was slightly more than 25,000, inevitably changes come with time. The twin lights on Thacher Island were now solar powered. Many people worked in an industrial park that was not maritime related, despite being named after Howard Blackburn.

But Gloucester had learned through its mistakes the importance of preserving the past. In 1966 the city had launched a program called "urban renewal," which some satirically called urban removal. Not unlike the British program that the people of Newlyn rebelled against in the 1930s, urban renewal was an attempt to modernize the waterfront. Historic treasures were torn down to make space for contemporary facilities. A house from the colonial period at the foot of Main Street was razed to make room for a much-needed gas station. The program raised an angry cry from some in Gloucester, led by poet Charles Olson. A few buildings were saved by this movement, including the granite Fitz Henry Lane house.

In the course of the urban renewal fight, Gloucester discovered that it was a city of people who loved it, and a great many of them were deeply concerned that, as the local expression goes, "Gloucester remain Gloucester." In the 1980s strict zoning regulations were established to preserve the waterfront, an area of varied and disorganized architecture and empty lots, much of it looking more like unplanned sprawl than a nearly 400-year-old city center. Now, downtown waterfront property could be used only for commercial marine activities. This not only barred condominiums and private homes but also the tourism industry. Hotels and yacht marinas were not permitted in the downtown waterfront. The idea was that someday the fishing would improve, but the industry would not be able to rebuild if it no longer had the waterfront. This fight over Gloucester's future comes up in local elections and dinner conversations. It is the central controversy of Gloucester.

Geoffrey Richon said, "As long as you keep the harbor industrial the ocean will always be something you can use. It will always represent commercial opportunities. Otherwise it will turn into Newport."

He could just as well have said that it would turn into Rockport. There, quarrying was finished, fishing was finished, and most of the boats in their harbor were for recreation. The red fishing shack in the downtown harbor no longer had a use but was carefully maintained because it was Motif #1. It made for more interesting paintings when it had been used and ramshackle and appeared as an element in a scene of a working fishing port. Tom Nicholas, one of Rockport's most celebrated painters, used it in the background of a painting of two lobstermen drinking coffee. But today the lobstermen are gone. Nicholas said, "When the motif becomes like a photograph, it loses what it could be—an incidental subject in a great painting."

Peg Williams, known for her colorful watercolors, grew up in Rockport. In an interview in 2006, when she was ninety-nine years old, she said, "Fishermen used to live in little shacks at the waterfront when I was a little girl and give me cookies. Now do you see the horrible things they are building there? Condos. I don't care, I'm on the way out."

Tom Nicholas's son, Tom Nicholas, Jr., also a painter living in Rockport, goes to Gloucester to paint. He said, "Growing up

I used to know fishermen and they had big cars and lots of money. Rockport's pretty much finished and there is less and less to paint in Gloucester Harbor. More and more I don't paint it the way it is there. I leave things out and add things. I get things from old photographs to put into paintings."

NOTHING BETTER ILLUSTRATED Gloucester's view of itself than the fight over the fate of a falling-down paint factory on the end of Rocky Neck, where boats pass from the outer to the inner harbor. The factory was built in 1870 by Tarr & Wonson. The Tarrs were one of the earliest families on Cape Ann and married into the Wonsons, one of the new families that arrived in 1716 after a fifth of the fishing fleet—five vessels and twenty men—was lost at sea.

In the 1860s, Tarr & Wonson became interested in an ancient maritime problem known as "fouling"—marine life attaching itself to the hull of ships, affecting both their longevity and navigation. In addition to destructive sea worms, one of the leading culprits are *Echeneis*, sometimes known as sucking fish because their anterior dorsal fin can be used as a suction cup to attach to objects. Enough of these fish attached to a hull will keep a ship from moving. Aristotle wrote about the problem in the fourth century B.C. To prevent fouling, the Phoenicians and Carthaginians covered their hulls with pitch. They may also have used copper sheathing. The Greeks used tar or wax. Lead sheathing was later used. Leonardo da Vinci designed a machine to roll out lead sheathing. Other ideas were a

second outer wooden hull; big-headed iron or copper nails
pounded in so close together that, in effect, they created a
sheathing; grease; sulphur; and oil.

The British navy built more ships with copper-sheathed
hulls, including two in 1789 that had all copper hulls and no
planking at all. But copper had its drawbacks, most notably
that it causes iron to corrode on contact. In 1824, Sir
Humphrey Davy, the pioneer chemist, worked on the problem
but was not able to solve it completely, so that once ships
started having iron hulls, copper became impossible to use. An
alternative to sheathing was paint. In 412 B.C., arsenic and sulphur were painted on hulls.

The paint factory

So when Tarr & Wonson started experimenting, they had a long history behind them. They arrived at a historic breakthrough formula of copper oxide in tar with naphtha or benzene. For a time the factory in Gloucester Harbor was the only copper-paint manufacturer in America. It was a huge commercial success, especially with vessels that went to the Caribbean, since in warm waters there is even greater fouling. Miami became one of the company's biggest markets. Jason Chamberlain, the great-great-grandson of H. Augustus Wonson, one of the company's founders, had worked in the factory as a youth and recalled:

In the center of the building there was a copper grinder that ran the full height of the building, and it was a stone grinder, and what would happen is they would drop copper nuggets half the size of a baseball roughly in the top of the thing, up on the third floor. With water pouring down on this grinding mill, which was two feet in diameter, the copper would be ground up by the stones gradually into smaller and smaller pieces until it came down the shaft out of the bottom in a powder form suitable for making it into paint.

Chamberlain remembered how it made "a huge racket." He also remembered that some of the workers had greenish skin. In the 1960s, the company started adding insecticides to the mix and also started making lead paint. Only toward the end were workers furnished with masks. In 1980 the entire factory was closed. What has been left is an abandoned toxic dump at the water's edge. But instead of a controversy about how to safely accomplish cleanup and removal, the fight, since the 1980s, has been about preservation. After the factory was closed, a plan to remove it was met with a public outcry. There was to be no more urban removal. This was a Gloucester historic site—part of the look of the harbor, in almost as many paintings as Motif #1 and just as red. And unlike the cleaned-up, rebuilt, fake shack in Rockport Harbor, this was the real thing, toxic and decaying—an authentic piece of Gloucester.

Proposals for a condominium or a private home raised a furor even after the developers promised to restore and maintain the original facade. One proposal was for Yale University to offer an art program from there. But is painting a marine industrial activity? That too had to be debated.

THE PAINT FACTORY was one of several prime waterfront properties, including two blocks of downtown harborfront, that had gone unused while their future and the future of Gloucester was debated. Richon said Gloucester was struggling to avoid "the Nantucketization of Cape Ann, where the postman and policemen fly in every day to do their work. If you keep enough to have a commercial port it will always be commercial. But you have to dredge the harbor and you have to have commercial docks. Once you let it go you can never get it back. We are in the right place. Close to the banks, far out into the North Atlantic. That's why Gloucester was built here."

But he recognized that change had to come and he was even

interested in hotel projects but not in the downtown water-front. "Things are changing and you are not going to be on an island that is different than everywhere else," he said.

Richon and investors bought the downtown shipyard from the Gloucester Marine Railway, the oldest marine railway in the United States. These were shipyards that hauled vessels up to dry dock on train tracks in high tide by a thick chain attached to an engine. The marine railway in Rocky Neck originally had a diesel locomotive that did the hauling. The locomotive is long gone, but both the Rocky Neck boatyard and the downtown one still haul boats out of the water on a railroad track. The two were owned by the same company, and in 1999 this old Gloucester maritime company was about to fail and close. It was a prime location in the heart of Gloucester Harbor, and Richon, fearing the location would lose its maritime function, gathered investors to buy it. Leaner and with an infusion of cash, the Rocky Neck yard was able to resolve its financial problems and prosper. The downtown marine railway yard, which agreed not to compete with the Rocky Neck company, restored old boats for historic uses. Among their projects was rebuilding the Novellos' 1936 side trawler, the *Vincie N*, into a reproduction of the *Eleanor*, one of three boats boarded in Boston Harbor in 1773 for the event known as the Boston Tea Party.

The Maritime Heritage Center was built next to the yard as an educational facility where children can learn about history and shipbuilding and marine biology. There is a small aquarium of local sea life, a marine biology lab, and a small museum.

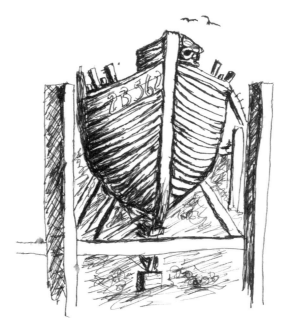

The Vincie N

Not all of Gloucester agreed with the preservationists. Mike Costello, a native of New England but not from Glouces-ter, was executive director of the Gloucester-based Cape Ann Chamber of Commerce. He said, "We don't want to hold these property owners hostage because the fish are coming back. This romantic dream, 'We don't want to be like Newport.' What's wrong with Newport?"

Jeff Worthly, who ran for mayor in 2005 and 2007 and was overwhelmingly defeated both times, said, "I look at the reality of the harbor. I see no reason why pleasure boats cannot dock next to fishing boats. The value of property goes up, city rev-enue goes up. If fishing returns they will outpay pleasure boats for space." But history shows that fishermen are rarely able to buy back space from yachtsmen.

These issues get fought over in every election and the preservation side has a history of garnering more votes. But Gloucester has a mayoral election every two years. Even within those two years the public can have a recall vote to remove the mayor, although this has been invoked only twice, in both cases unsuccessfully. There are usually numerous candidates for mayor, and there is a sense that no one knows who is going to run next. To be pro-preservation means to be pro-fishery, and this is a potent stand but it has never translated into political power for fishermen themselves. In the 2007 mayoral race Carolyn Kirk, a business consultant who knew almost nothing about fisheries, soundly defeated James Destino, a former fisherman. Kirk convinced voters in the near-bankrupt city that she was a better financial manager.

One aspect of the island mentality that has remained in place is that in local elections, who is a Republican and who is a Democrat is a topic that is rarely mentioned. Such off-island politics is considered irrelevant. Some candidacies are short-lived. Salvatore Frontiero, a retired 78-year-old fisherman, announced his candidacy for mayor in February 2007. He billed himself as a "problem solver" and said "I would like to help." But the following day while planning his campaign he drifted off to sleep, and in that sleep his wife, who had died ten months earlier, came to him and warned him that being mayor would kill him. "My dreams never fail me," said Frontiero, announcing his withdrawal. On the same day the chances of another candidate, Kathleen Ann Martin, were considerably diminished when she was arrested for disorderly conduct in the local library.

The inevitable candidate is Dan Ruberti, who has run for mayor every two years since 1975, when he replaced an earlier perennial candidate—a junk man named Clancy. Clancy the Junkman once came close enough to winning to frighten half of Gloucester. Ruberti refused to finance his campaigns, calling himself the originator of campaign reform, but he stood on street corners passing out handwritten bumper stickers that said, "Stem the tide."

"Stem the tide" is a nautical term, he explained, for going forward against the tide. "Leadership is going forward against all obstacles." His previous slogans included "Isn't it high tide we vote for Ruberi?" and "A lobster in every pot." He once came in second and lost with 28 percent against a five-time incumbent.

Ruberti's house on the western outskirts of town was known by most locals because it looked like the permanent site of a yard sale, with bicycles, boxes, crates, more than one hundred—according to him—Tonka Trucks, scrap metal, hoses, and faucets. Picking up some of the scrap plumbing he said reflectively, "An artist could do something with this." His basement was stacked with boxes of Boston Celtics T-shirts and his stairway was jammed with boxes filled with carefully banded used Massachusetts lottery tickets. "Every now and then there is an unclaimed winning ticket," he enthusiastically explained. "It could be here."

"I am a down-to-earth person," stated Ruberti, a pleasant, smiling septuagenarian. Like most mayoral candidates in

Gloucester, he was passionate about the city. "I ran for mayor because I felt I would be very capable. I'd work my ass off."

"But you'll never be mayor," his wife, Pat, added. "For two reasons. You don't take any money and you don't get involved in politics in off years."

Dan slapped the table as though this were a great joke and countered, "They may say someday, 'Let's give the son-of-a-gun the job!' "

His wife confessed that she would prefer to vote for other candidates, "but Dan wouldn't let me."

Ruberti was not listening. He was fumbling through boxes of cheap costume jewelry that he found in people's garbage. He tossed the worthless pieces to the side but kept looking, confident a gem would turn up. "Damn it," he said. "I'm the best thing since peanut butter. I've proven myself to myself but I want to prove it to everybody else. I am very ambidextrous and tough as nails."

It was not always noticed that Ruberti raised some good issues. He kept warning that "the kids who are born here will have no chance of ever owning a home here." He was the town Chicken Little. For years he said that the tower in City Hall would fall. In his yard he collected pieces of slate that had fallen from the tower. "This building will implode," Ruberti said in each campaign. "Wait till you hear the sound when the bell falls. I warned them that the tower was tilting but everyone said it's just that crazy Dan Ruberti."

Once when John Bell was mayor, a City Hall meeting was interrupted by a piece of the tower crashing to the ground. Bell sometimes found pieces that had landed in the balcony of his office. By 2007 the city recognized that the most important landmark in Gloucester, the clock tower on the 1871 City Hall, a dome that defined the distinctive skyline of Gloucester Harbor and is featured in many paintings, was dangerously tilting. Some $150,000 was spent on netting the tower to catch falling debris, but the city was in search of more than $1 million to repair the landmark.

The truth was that the city of Gloucester was desperately short of income. This became tragically apparent in 2006 when a woman died in a fire two miles from a fire station that had been closed by budget cuts. In 2006 the state ordered the city to spend tens of millions of dollars that it didn't have to improve the inadequate sewer system. The Gloucester school system spent about a half million dollars more than was in their budget. The city's financial records were also deemed disastrous. The *Daily Times* wrote, "They may not have much money but they don't know where it is."

Meanwhile, as real estate prices were rising, so was poverty. A 2006 study by a nonprofit organization, Project Bread, reported that the number of Gloucester families that skipped meals because they could not afford food had doubled in three years.

These problems are due as much to politics as to lack of income. From 2002 to 2006, under Governor Mitt Romney, state aid to Gloucester was decreased by more than $20 million, at the same time as the federal government was placing more demands on municipalities. But all this aside, it was clear that Gloucester needed to take in more money, and the obvious choice, as with all coastal towns, was to embrace that booming and destructive force, the tourism industry.

Gloucester has a long tradition of "summer people," but far less of the tourist who comes for a day or two. Its restaurant offerings are notoriously rustic, with a few exceptions, and the clam shacks are better in Essex. Gloucester somewhat proudly lacks the better hotels and charming New England bed-and-breakfasts of Rockport. It has some, but surprisingly few, shops, boutiques, and galleries. Tourists stay in Rockport and visit Gloucester.

But then came the phenomenon known as "the Perfect Storm." On Halloween in 1991 a storm blew through that the *Gloucester Daily Times* labeled "an unexpectedly potent nor'easter." Nor'easters—storms that blow in from the northeast of long duration with destructive high winds and black skies—are to coastal New England what tornadoes are to the Plains.

This nor'easter was so powerful that it demolished dozens of Cape Ann homes and even budged the twenty-ton granite boulders along the coast. The *Andrea Gail*, a Gloucester swordfish boat, had left the week before with a Gloucester captain, Billy Tyne, and a five-man crew, two of whom were from Gloucester. They were fishing the Grand Banks off of Newfoundland beyond the 200-mile limit and were thought to have been some seven hundred miles northeast of Gloucester when the storm hit. Tyne radioed that he was in thirty-foot seas with fifty- to eighty-knot winds. He was never heard from again. Neither the vessel nor any of the six crewmen were ever found.

The *Andrea Gail* was the only Gloucester vessel lost in this storm, which given Gloucester's history was not a bad storm— just one more Gloucester story. But in modern times with modern communications and weather forecasting, though tragedies still regularly befall fishermen, there are not the sweeping disasters of past centuries. The *Andrea Gail* was the first Gloucester vessel to be lost with all hands since 1978, when a dragger and a swordfishing boat were both lost with nine men. As with most fishing boats lost at sea, no one knows what happened to the *Andrea Gail*. That is part of the agony of this type of loss. And as familiar as this kind of tragedy is in Gloucester, the news was painful to fishermen, their families, the community, and the numerous people who had lost other friends and relatives in the past. It always is. Roger Nowell, a Newlyn fisherman, said, "Every time you hear about another boat going down, it always reminds you of your own mortality."

In 1997, Sebastian Junger, a young author, published *The Perfect Storm*, an account of the fate of the *Andrea Gail*. Though written as a nonfiction book, much of it was imagined, of necessity, because the events were unknowable. In Gloucester, *The Perfect Storm* had a number of strikes against it. Junger was not from Gloucester but from an affluent Boston suburb. He had even less fishing experience than Rudyard Kipling, by his own admission, never having been at sea on a commercial fishing vessel. Gloucester people found numerous flaws in the account, and at the outset there was a great deal of grumbling and even threats of lawsuits. But the book sold millions. They liked that in Gloucester. Better yet, there was a Hollywood version and it was shot in Gloucester.

Films had been made in Gloucester before. In 1919, Salt Island was used for the set of a Fox Films silent thriller called *Bride 13*. Fox built a huge medieval castle out of wood and plaster and the locals were kept amused at the ill-fated rescue of the heroine when a stuntman was blown off course in a hot air balloon, and then the failed attempt at blowing up the castle, and, even better, the successful one when the entire structure vanished in a noisy smoke-filled second. There was *Captains Courageous* in 1937, starring Spencer Tracy, with a closing shot of the *Man at the Wheel*. In 1941, Gloucester was mentioned in *The Man Who Came to Dinner* as the site of an axe murder that had actually taken place in Fall River. But it had been a long time since Hollywood had come to Gloucester to tell a real Gloucester story.

The movie *The Perfect Storm* was not the blockbuster hit that Warner Brothers had hoped for, but suddenly people started showing up in Gloucester. They wanted to know where the Crow's Nest was, because it was in the movie. The Crow's Nest was, even by Gloucester standards, a notoriously seedy waterfront bar. It was so small and cramped that the film crew was unable to shoot in it and so they reconstructed it on the harbor, which is where many of the tourists later turned up looking for it. In the movie, one actor wore a T-shirt from the local ice company, Cape Pond Ice. Suddenly the ice company that was founded in Gloucester in 1896 had a lucrative T-shirt trade.

Locals started earning a living by giving "*Perfect Storm* tours" of Gloucester. The real Crow's Nest, avoided by most locals, became a tourist attraction. There a tourist could see authentic down-and-out Gloucester alcoholics drinking. A shop appropriately named "The Tourist Trap" was opened next door, selling *Perfect Storm* souvenirs such as Cape Pond Ice T-shirts. Sam Novello had contemplated taking tourists out for *Perfect Storm* adventures. "It would be like a dude ranch," he said. He even thought he could simulate mini-squall conditions to give them an authentic experience.

The tourists wanted to see the monument to the fishermen lost at sea, and it seemed a good touch to put up plaques listing the victims. The only such list in existence was at the City Hall, and it listed fishermen from Gloucester who were lost at sea. Half the crew of the *Andrea Gail*—the two from Florida and the one New Yorker—would not qualify for such a list. Instead, this new series of plaques listed the fishermen lost at sea who had shipped out of Gloucester. Exhaustive examination of the city's archives turned up 5,368 names—more than a thousand more than had been on the City Hall list, but still perhaps only half as many as historians think the real number may be. The names of many of the fishermen who went down in the early years are unknown.

In 2007, The Tourist Trap closed for lack of business. *Perfect Storm* tourism had faded away. Many Gloucester people would like to replace it. Gloucester is not opposed to tourism as long as it doesn't intrude on the fishermen's space. In recent years they have been trying to attract cruise ships. For Gloucester, and for all struggling coastal towns, the question remains unanswered: How can the community cash in on the tourism boom without destroying the character of the town?

GLOUCESTER PEOPLE love Gloucester. Those who leave often come back. In 1946 Charles Olson, returning to Gloucester, wrote to Ezra Pound, "This is home. I got here, and thought to turn right back. But the place, the sea, and the wide light which it also has in October won't let Connie and me go."

Gloucester has always been about fishing, and to many it would be tragic to lose it. "I cannot see the fishing industry dying," said Gloucester's nonagenarian Poet Laureate, Vincent Ferrini. "I won't permit myself to think it."

Fishing remains extremely dangerous, though it is far safer than it once was. Fishermen now have survival suits of insulated waterproof flotation material. They also have cell phones and long-range satellite phones, and when their boat goes down they can sit in an inflatable raft and give the Coast Guard their position—assuming the seas are not too high for a raft to survive. Angela Sanfilippo, whose husband, John, was rescued at sea, said, "I cannot understand this romance about the great days of sail. They lost so many ships, so many men. Now a storm comes and the Coast Guard calls me and says get your husband home." When John is at sea they are constantly in touch by cell phone. "Cell phones," she said, "are a fisherman's best friend."

Today boats are lost and the fishermen are saved, as happened to Joe Santapaola's sons when their vessel, the *Melon II*, collided with another and sank in 2005. The same year a scal-

Chopping ice from the trawler Saturn, *January 28, 1928*
(COURTESY OF THE CAPE ANN HISTORICAL ASSOCIATION,
GLOUCESTER, MASSACHUSETTS)

lop boat sank off Nantucket when the dredger caught on the bottom: the captain was lost but the two crewmen, one from Gloucester, badly dehydrated, were rescued from a life raft.

Fishermen are still dying from some of the same things that sunk the old schooners. If too much ice builds up on the windward side on a winter trip, the boat becomes unstable and can capsize. In 1980 a fishing boat sank in a sudden squall off Eastern Point, right in the harbor, and one of the crew drowned in sight of people eating lunch in Eastern Point homes with their sea view. In 2007 a dead fisherman was found floating in the harbor.

When Sam Novello was a boy his father asked him what he wanted to do with his life. "I said, 'I don't know,' and he said, 'Then you're going fishing.' " Novello now says, "It was a good life. You were free. I just wanted to make a living and pay my bills and be free."

Fishermen defied the dangers because they had a life like none other. They were independent, working for themselves, with no one telling them what to do. That is not true anymore. Now when a fisherman leaves the dock he calls in to regulators to

report that he is leaving and the clock on his days-at-sea al-
lowance starts running. Sometimes there is an observer on
board to monitor his activities. Even if not, an electronic device
called a Vessel Maintainance System Transmitter enables regulators to monitor the vessel's every movement. The fisherman
is required to report where he is fishing and the species he is
targeting and the size of the catch, and must arrange to have
an official meet him at the dock when he comes in.

A lot of fishermen have left fishing, saying it just isn't fun
anymore. In 2006 ex-fisherman Rich Arnold said, "I enjoyed
fishing every moment, just loved it until four or five years ago.
I quit in 2000 . . . Filling out logs and measuring every fish. It
was just plain aggravation." The stores and businesses of
downtown Gloucester are increasingly being staffed with ex-
fishermen.

Vito Giacalone, still burly and fit in his mid-forties, has the
kind of confident physical presence worn by great athletes and
astronauts. He had gotten out of fishing, the family trade, and
had moved into Gloucester's new growth industry, real estate.
But in 2006 he bought *The Lady Grace*, a movie star because it
had been used to play the ill-fated *Andrea Gail* in the film of
The Perfect Storm. He was refitting it as a bottom dragger.
Asked why he would go back to fishing now with everything
that was going on, he smiled and said, "I love it." His smile
turned a little mischievous. "I always said I would go back if I
didn't have to make a living at it. Now you can't."

Most in the fishing industry are confident that the stocks
will be back, at least some of the stocks. People who have seen
this much tragedy aren't alarmists. But they assume that the

fishing will be on a smaller scale. Butch Maniscalco of the
Gloucester Display Fish Auction said, "I remember the volume
I did with my father. The thousands of pounds." Everyone has
those reminiscences and no one expects those days to be back.
But they do believe that if they can hang on to the waterfront,
the ones who survive will do well. "It's going to work out all
right," Gus Foote said. "We've had hard times before."

Could this be true? Or is this the last fish tale? Will the fishermen, the regulators, and the scientists out of their raucous
brawl eventually find the right formula or will Gloucester's
slow descent into another seaside resort be the final Gloucester
story? The nature of coastal society is changing in ways it
never has in this country for centuries, and in Europe for mil-
lennia. And Gloucester, Newlyn, and a handful of other ports on
every coastline of every sea fight on, their future uncertain. The
consequences to the planet of the destruction of ocean life are
terrifying. Scientists are increasingly worried by the loss of biodiversity. As more and more varieties of life disappear, it becomes increasingly difficult for the planet to sustain any life.

Next to that calamitous scenario, the survival of the few remaining picturesque fishing towns may seem like a small thing.
But intertwined with the issue of biodiversity is the idea of
sociodiversity—social diversity. Each culture, each way of life
that vanishes diminishes the richness of civilization and makes
it more difficult for civilization to prosper. The multiplicity of
cultures, like the multiplicity of biological species, is the guaran-
tor of the continuation of life. The Mexican poet, Octavio Paz,
once wrote, "Every view of the world that becomes extinct,
every culture that disappears, diminishes the possibility of life."

It may be that the reason the fishing life, the life of Gloucester, will go on is that fishermen are willing to make the necessary sacrifices. For it to be over, for New England and all the coastlines of the world to house nothing but tourism and yachting, for Tragabigzanda to end after all these centuries, for Gloucester to no longer be Gloucester, would be, as the Poet Laureate Ferrini said, unthinkable.

Lone fishing boat passing Good Harbor Beach on her way home

Bibliography

CAPE ANN

Babson, John J. *History of the Town of Gloucester, Cape Ann.* Gloucester: Peter Smith, 1972.

Babson, Roger W., and Foster H. Saville. *Cape Ann Tourist's Guide.* Gloucester: Cape Ann Community League, 1946.

Connolly, James B. *The Port of Gloucester.* New York: Doubleday, Doran & Company, 1940.

Cooley, John L. *Rockport Sketch Book.* Rockport: Rockport Art Association, 1965.

Deennen, W. H. *The Rocks of Cape Ann.* Gloucester: Gloucester Cultural Commission.

Dresser, Thomas. *Dogtown: A Village Lost In Time.* Franconia, N.H.: Thorn Books, 2001.

Dunlap, Sarah V. *The Jewish Community of Cape Ann: An Oral History.* Gloucester: Cape Ann Jewish Community Oral History Project, 1998.

Erkkila, Barbara H. *Hammers on Stones: The History of Cape Ann Granite.* Gloucester: Peter Smith, 1987.

Fifield, Jr., Charles Woodbury. *Along the Gloucester Waterfront 1938 to 1946.* Gloucester: Cape Ann Ticket & Label Company, 1955.

Garland, Joseph E. *Eastern Point: A Nautical, Rustical, and More or Less Sociable Chronicle of Gloucester's Outer Shield and Inner Sanctum.* Beverly: Commonwealth Editions, 1999.

———. *The Gloucester Guide: A Stroll through Place and Time.* Charleston, S.C.: The History Press, 2004.

————. *Boston's Gold Coast: The North Shore 1890–1929.* Boston: Little Brown, 1981.

————. *Lone Voyager: The Extraordinary Adventures of Howard Blackburn, Hero Fisherman of Gloucester.* New York: Touchstone Books, 2000.

Hammond, John Hays. *The Autobiography of John Hays Hammond,* two volumes. New York: Farrar & Rinehart, 1935.

Heyerman, Christine Leigh. *Commerce and Culture: The Maritime Communities of Colonial Massachusetts 1690–1750.* New York: W. W. Norton, 1984.

Oaks, Martha. *The Gloucester Fishermen's Institute 1891–1991.* Gloucester: Fishermen's Institute, 1991.

O'Gorman, James F. *This Other Gloucester.* Boston: James F. O'Gorman, 1976.

Parsons, Eleanor C. *Thachers: Island of the Twin Lights.* West Kennebunk, Maine: Phoenix Publishing, 2000.

Pollack, Susan. *Gloucester Fishermen's Wives Cookbook: Stories and Recipes.* Rockport: Twin Lights, 2005.

Procter, George H. *The Fishermen's Memorial Record Book.* Charleston, S.C.: The History Press, 2004 (original 1873).

Ray, Mary. *Gloucester, Massachusetts: Historical Time Line 1000–1999.* Gloucester: Gloucester Archives Committee, 2002.

Story, Dana. *Frame Up! A Story of Essex, Its Shipyards and Its People.* Charleston, S.C.: The History Press, 2004.

————. *Growing Up in a Shipyard: Reminiscences of a Shipbuilding Life in Essex, Massachusetts.* Mystic: Seaport Museum, 1991.

Sucholeiki, Irving. *A Return to Dogtown: A Look at the Artifacts Left Behind by Some of Cape Ann's Early Settlers.* Gloucester: Chisholm and Hunt Printers, Inc., 1992.

The Fisheries of Gloucester from 1623 to 1876. Gloucester: Procter Brothers Publishing, 1876.

St. Peter's Fiesta Through the Years. Gloucester: The Young Men's Coalition, 2001.

GENERAL

Bradford, William. *Of Plymouth Plantation.* New York: Alfred A. Knopf, 2001.

Clover, Charles. *The End of the Line: How Overfishing Is Changing the World and What We Eat.* New York: The New Press, 2006.

Goode, George Brown. *The Fisheries and Fishery Industries of the United States,* seven volumes. Washington, D.C.: Government Printing Office, 1884–87.

Graham, Michael. *The Fish Gate.* London: Faber & Faber, 1943.

Hooker, Richard J. *The Book of Chowder.* Boston: The Harvard Common Press, 1978.

Innis, Harold A. *The Cod Fisheries: The History of an International Economy.* Toronto: University of Toronto Press, 1954.

Kupperman, Karen Ordahl, ed. *Captain John Smith: A Select Edition of His Writings.* Chapel Hill: University of North Carolina Press, 1988.

Kurlansky, Mark. *Cod: A Biography of the Fish That Changed the World.* New York: Walker, 1997.

Morey, George. *The North Sea.* London: Frederick Muller, 1968.

Nowell, Roger, and Jeremy Mills. *The Skipper: A Fisherman's Tale.* London: BBC Books, 1993.

Perry, Margaret E. *Newlyn: A Brief History.* Penzance: Margaret Perry, 1999.

———. *Mousehole: A Brief History.* Penzance: Margaret Perry, 1998.

Preble, Dave. *The Fishes of the Sea: Commercial and Sport Fishing in New England.* Dobbs Ferry, N.Y.: Sheridan House, 2001.

Sagar-Fenton, Michael. *About Penzance, Newlyn & Mousehole.* Launceston, Cornwall: Bossiney Books, 2000.

Trewin, Carol. *Cornish Fishing and Seafood.* Penzance: Alison Hodge, 2006.

Van der Merwe, Pieter, ed. *Hooking, Drifting and Trawling: 500 Years of British Deep Sea Fishing.* London, National Maritime Museum, 1986.

Wigan, Michael. *The Last of the Hunter Gatherers: Fisheries Crisis at Sea.* Shrewsbery: Swan Hill Press, 1998.

Woods Hole Oceanographic Institute. *Marine Fouling and Its Prevention.* Menasha, Wisc.: George Banta Publishing, 1952.

ART

Davies, Kristian. *Artists of Cape Ann: A 150-Year Tradition.* Rockport: Twin Lights, 2001.

Elleman, Barbara. *Virginia Lee Burton: A Life in Art.* Boston: Houghton Mifflin, 2002.

Laity, John Curnow. *Newlyn Copper.* Penzance: Morrab Studio, 1998.

Taylor, Sue, ed. *Winslow Homer in Gloucester.* Chicago: Terra Museum of American Art, 1990.

Wallace, Catherine. *Under the Open Sky: The Paintings of the Newlyn and Lamorna Artists 1880–1940, in the Public Collections of Cornwall and Plymouth.* Truro, Cornwall: Truran, 2002.

Wilkin, Karen. *Stuart Davis in Gloucester.* West Stockbridge, Mass.: Hard Press, 1999.

Williams, Douglas, and Paul Bodmin. *Around Newlyn, Mousehole.* Cornwall: Bossiney Books, 1988.

Wilmerding, John. *Fitz Henry Lane: A reprint of a 1971 book entitled "Fitz Hugh Lane" with new information on the artist's name.* Gloucester: Cape Ann Historical Association, 2005.

Folly Cove Designers. Gloucester: Cape Ann Historical Association, 1996.

George Demetrios: Sculptor and Teacher. Gloucester: Cape Ann Historical Association, 1986.

Marsden Hartley: Soliloquy in Dogtown. Gloucester: Cape Ann Historical Association, 1985.

Masterworks. Gloucester: North Shore Art Association, 1997.

GLOUCESTER LITERATURE

Allen, Donald, and Benjamin Friedlander, eds. *Collected Prose: Charles Olson.* Berkeley: University of California Press, 1997.

Butterick, George F., ed. *The Maximus Poems: Charles Olson.* Berkeley: University of California Press, 1975.

Clark, Tom. *Charles Olson: The Allegory of a Poet's Life.* Berkeley: North Atlantic Books, 2000.

Creeley, Robert. *Charles Olson: Selected Writings.* New York: New Directions, 1951.

Eliot, T. S. *The Wasteland.* Manuscript with annotations by Ezra Pound. New York Public Library, Berg Collection.

Ferrini, Vincent. *The Autobiography of Vincent Ferrini.* Gloucester: Ten Pound Island Press, 1988.

Horovitz, Israel. *New England Blue: 6 Plays of Working-class Life.* Lyme, N.H.: Smith and Krauss, 1995.

Kipling, Rudyard. *Captains Courageous.* Pleasantville, N.Y.: Readers Digest, 1994.

Maud, Ralph, ed. *Selected Letters: Charles Olson.* Berkeley: University of California Press, 2000.

ARTICLES

Burbank, Russell P. "Foreign Fish Kill Birdseye Birthplace." *Boston Globe,* July 18, 1965.

Greer, Dave. "When Salt Barks Sailed from Sicily to Gloucester." *Cape Ann Summer Sun,* July 6, 1950.

Howard, Robert West. "One Jump Ahead." *Nation's Business,* July 1949.

Morris, Steven. "Skippers in the Dock as Cornwall's Last Great Fishing Town Awaits Fate." *The Guardian,* London, August 28, 2007.

Smylie, Mike. "Newlyn—A Fishing Port on the Western Approaches." *Maritime Life and Traditions,* Winter 2004.

Whitehill, Walter Muir. "Gloucester Has Been a Maritime Community for Three Centuries." *Gloucester Daily Times,* August 18, 1942.

Worm, Boria, Edward B. Babier, Nicole Beaumont, J. Emmett Duffy, Carl Folke, Benjamin S. Halpern, Jeremy B. C. Jackson, Heike K. Lotze, Fiorenza Micheli, Stepjen R. Palumbi, Enric Sala, Kimberley A. Selkoe, John J. Stachowicz, Reg Watson. "Impacts of Biodiversity Loss on Ocean Ecosystem Services." *Science,* 314 (2006).

Gloucester Telegraph, January 4, 1862; March 19, 1862.

"An American Fishing Port." *Lippincott's Magazine,* May 1868.

"Appointed Committee to Crystalize Feelings at Other Ports," *Gloucester Daily Times,* October 3, 1911.

ORAL HISTORY ARCHIVES

Interview with Manuel P. Domingo, in Gloucester, February 21, 1978; interviewed by Linda Brayton and David Masters for Toward an Oral History of Cape Ann.

Interview with Joseph Garland, in Gloucester, August 13, 1978; interviewed by Linda Brayton and David Masters for Toward an Oral History of Cape Ann.

Interview with Walker Hancock, in Gloucester, July 22, 1977; interviewed by Robert Brown for the Archives of American Art, Smithsonian Institute.

Interview with Samuel Hershey, March 1978; interviewed by Linda Brayton and David Masters for Toward an Oral History of Cape Ann.

Interview with William Meyerowitz and Theresa Bernstein, in Gloucester, June 23, 1978; interviewed by Linda Brayton and David Masters for Toward an Oral History of Cape Ann.

Interview with Lena Novello, in Gloucester, January 21, 1978; interviewed by Linda Brayton and David Masters for Toward an Oral History of Cape Ann.

Acknowledgments

My thanks to the city of Gloucester; to Larry Swift and Marilyn Swift, the wondrous watercolorist, who have been such good friends and showed us so much; to the Hrubys, Ken and Billie, most especially Melissa Palladino and all of the Hruby family for warmth, friendship, and support; to John Bell and especially to Janis Bell whose help, support, and kindness has made all the difference; to my friend Joseph Garland, who has done more than anyone to document the history of Gloucester and who generously shared his stories, his staggering knowledge, his great decency, and his wicked sense of humor; to all of the staff of The Book Store, the Sawyer Free Library, and the Cape Ann Historical Museum, and to my old friend, the tough and cozy old town of Newlyn; to Margaret Perry, another great local historian—its another thing that is nurtured by fishing cultures—for all her help and kindness in Newlyn. And to my great friend and superb British editor, Will Sulkin, for his help and support.

This book owes an important debt to my editor, Nancy Miller, one of the last of the great, and to my agent and friend Charlotte Sheedy, and also to Lea Beresford and Meredith Kaffel who saved me in a thousand different ways.

And lastly, to Marian and Talia with thanks and more love than I could ever write, for sharing with me the Tragabigzanda extravaganza, and that foaming, rock-studded coast that always grabs my heart.

Index

Page numbers in *italics* refer to illustrations.